SCIENCE 101

WEATHER

SCIENCE 101: WEATHER. Copyright © 2007 by HarperCollins Publishers.
All rights reserved. Printed in the United States of America. No part of
this book may be used or reproduced in any manner whatsoever without
written permission except in the case of brief quotations embodied in critical
articles and reviews. For information, address HarperCollins Publishers,
10 East 53rd Street, New York, NY 10022.

HarperCollins books may be purchased for educational, business, or sales
promotional use. For information, please write: Special Markets Department,
HarperCollins Publishers, 10 East 53rd Street, New York, NY 10022.

Produced for HarperCollins by:

Hydra Publishing
129 Main Street
Irvington, NY 10533
www.hylaspublishing.com

FIRST EDITION

The name of the "Smithsonian," "Smithsonian Institution," and the
sunburst logo are registered trademarks of the Smithsonian Institution.

Library of Congress Cataloging-in-Publication Data has been applied for.

ISBN: 978-0-06-089137-4
ISBN-10: 0-06-089137-8

07   08   09   10   QW   10   9   8   7   6   5   4   3   2   1

SCIENCE 101

WEATHER

Trudy E. Bell

Collins
An Imprint of HarperCollins*Publishers*

CONTENTS

WELCOME TO WEATHER

Left: Cloud-to-ground lightning brightens the night in Tucson, Arizona. Top: Sunset at Beausoleil Island in Ontario, Canada. The variable quality of weather and clouds ensures an ever-changing display when the Sun sets. Bottom: Frost-coated leaves, photographed on an island in Puget Sound, Washington.

Step outside any time, dawn or dusk, day or night. Is it clear or cloudy? Warm or cold? Stormy or calm? Is there dew or frost on the grass, a halo encircling the moon, or a rainbow arching across the sky?

Behold the weather!

As adults, most of us take weather for granted, often just glancing out the window to see whether to grab an umbrella before heading off to work in a hermetically sealed automobile, bus, or train. It seems as if the only time busy people pay attention to the weather is if they plan to be golfing, boating, or skiing on the weekend and they "need" it to be sunny or to snow—or if an alert interrupts a television or radio program to announce a tornado watch because of the approach of a severe storm with dangerous hail.

Yet, while walking across that parking lot from car to office, stop a moment. Look up, and truly see the sky.

WEATHER APPRECIATION

The marvelous thing about weather is that anyone of any age can appreciate it from any location at any time, with absolutely no equipment—just open eyes and mind. Indeed, that is exactly how the earliest weather pioneers in the seventeenth, eighteenth, nineteenth, and even the twentieth centuries made many discoveries, or at least became inspired by intriguing questions.

With your own senses, you can often detect what TV weather personalities are pointing out on satellite maps. If they mention that a cold front is moving into the area, step outside: See if there's a brisk breeze and the temperature and humidity are dropping. Look up: See half the sky thick with clouds, and the other half sparkling clear and blue, the clear air seeming to push away the clouds in front of it.

Stay alert for delightful surprises: Have you sometimes seen unexpected patches of rainbow-hued colors in clouds, or what looks like the arc of a rainbow itself directly overhead, even when there has been no rain? Most likely, you are seeing sunlight broken into its spectrum of colors by clouds of ice crystals. Indeed, such ice-crystal displays, although famed for their rarity, are actually more common than real rainbows.

Even gray days hold a fascination, especially if the air is thick and damp with fog. Watch how fog moisture collects on leaves and drips to the ground: Indeed, in some climates, such as in the redwood forests along the California coast, fog moisture actually makes up most of the annual precipitation.

Go further. Ask questions of what you see. Do the clouds aloft seem to be suspended at different levels? It's true: Clouds generally do cluster at three different heights in the atmosphere, and sometimes you can see all three levels at once. For a few tens of dollars, build a weather station out of household items to mount in your yard (see page 205), augmenting your senses by measuring rainfall, air pressure, and humidity with surprising accuracy—and compare your own local short-range weather predictions with what you observe.

SCIENTISTS ALOFT

Today, meteorology is high-tech science, composed of equal parts observation, theory, and high-speed supercomputing. Ground-based

Rainbows can often appear after intense thunderstorms, such as this full arc photographed in rural central Illinois.

The aurora borealis shimmers above a Canadian forest. The spectacular lights arise when charged particles from the solar wind collide with Earth's magnetic field and upper atmosphere.

weather radars track the movement and nature of thunderstorms and other precipitation, thousands of automated weather stations regularly radio their readings to central receivers, weather balloons bear instruments high into the atmosphere, and satellites monitor large weather patterns over much of the globe.

Meteorology can also be an adventure science. Instead of avoiding the dangers of extreme weather conditions, storm chasers and hurricane hunters risk their lives to drive or fly very near or even right into storm systems, gathering data and photographs unobtainable any other way.

Measurements from these and other sources are fed into atmospheric general circulation models—huge number-crunching supercomputer simulations—to see how well scientific theory and empirical observations combine to predict outcomes. The ultimate goal: to calculate truly long-range weather forecasts an entire season or even several seasons in advance. Eventually, such predictions may answer such questions as: Will the winter be good for ski resorts? Will the summer be bad for growing corn? Will this hurricane season be dire or just dangerous?

WEATHER AND CLIMATE

Weather is here today and different tomorrow, but climate is a big-picture view of long-term weather, both for individual regions and for the world as a whole.

Human industrialization affects weather, with deforestation leading to changes in regional patterns of precipitation. Air pollution from power plants, factories, and automobiles in middle latitudes leads to acid precipitation as far away as the Arctic or the stratosphere over Antarctica. Greenhouse gas emissions may be changing the global climate in ways we have yet to understand. Perhaps the biggest question facing society today is whether or not the world's climate is changing as a result of human activities, and if so, what humans can, or should, do.

A scientist collects samples from ancient glacial ice in West Antarctica to test for evidence of atmospheric greenhouse gases.

3

CHAPTER 1

AN OCEAN OF AIR

Left: Cumulus clouds over mountainous terrain in Estepona, Spain. All life on Earth is shaped by forces of weather, our first and closest link to the atmosphere that surrounds our planet. Top: The Algodones Dunes, a desert area in Southern California, near the border of Arizona. Bottom: Lightning strikes during a storm in a tropical region of Africa. These widely varying climates exemplify weather's tremendous impact on Earth, its vegetation, and the life it sustains.

Regardless of age or profession, location or wealth, all humans breathe the air. One can survive more than a month without food and even several days without water, but four short minutes without oxygen brings death. People even bottle the atmosphere to carry along when they dive into the oceans, climb the highest mountain peaks, or venture into outer space.

All life on Earth—people, plants, and animals—lives at the bottom of a vast ocean of air called our atmosphere. Although composed of gases, the atmosphere has distinct layers, like the ocean. It even has currents: jet streams and other prevailing winds. The atmosphere also interacts intricately with both the ocean and land masses.

Compared with the radius of the Earth (about 3,963 miles or 6,378 km), the atmosphere is quite shallow. Yet this thin layer of gases shields Earth from high-speed celestial rocks (meteorites) and harmful radiation, and provides oxygen for breathing—making life possible on this planet.

The atmosphere is also home to Earth's weather. Weather wild and majestic shapes our planet. Although we can understand or predict the weather, we still can neither tame nor control it. Farmers still eagerly beseech the sky for rain, and coastal residents in hurricane areas still tape their windows and evacuate their homes. And all humans pause to stand in awe before a brilliant rainbow or glorious sunset.

Layer-Cake Atmosphere

With all the winds that sweep across land and sea, you'd think that the atmosphere would be all mixed up and of fairly homogenous composition.

But you'd be wrong. Earth's atmosphere is highly structured. In fact, it can be viewed as a five-layer cake hundreds of miles thick. Each layer has its own distinct chemical composition, movement, density, and changes in temperature. Each layer of the atmosphere is bounded by "pauses" where the changes in its physical properties are most abrupt.

How thick is the atmosphere? There seem to be several answers. Fully 99 percent of the atmosphere's mass and moisture are in the two lowest layers; together, these layers are only some 30 miles (48 km) thick. Scientists differ on where the nominal edge of outer space begins, placing it between 50 and 62 miles' (about 80–100 km) altitude—a region where meteors are still incinerated by outlying atmospheric molecules. Many spacecraft in low Earth orbit 150 to 300 miles (240–480 km) high are slowed by the friction of air molecules rubbing against them as they pass through the tenuous

Right: Artist's rendition of the meteor-detecting Pegasus satellite, in low Earth orbit in 1965. Top left: The Sun's hot outer corona, shown during an eclipse.

uppermost layers of the atmosphere where auroras (northern and southern lights) flicker and dance. In fact, atmospheric friction is so significant that some spacecraft have spiraled down to Earth unless they are periodically reboosted to the desired altitude.

TEMPERATURE VS. HEAT

Most people equate temperature with the heat or cold they feel—a winter's day with temperatures below freezing is frigid, whereas a summer's day with temperatures topping 100°F (38°C) is sweltering. But in meteorology and other sciences, the definition of "temperature" differs from common usage. That difference is essential to understanding discussions about the atmosphere, the oceans, and climate.

Temperature and heat are related, but they are not the same thing. A burning match has a much higher temperature than a steam radiator, but the match can't heat a living room. Boiling water just poured from a teakettle to make a cup of tea has the same temperature as the water in the kettle, but more ice cubes could be melted in the kettle than in the teacup because the kettle is so much bigger than a cup. Thus, something small at a high temperature may not give off much heat. Conversely, something massive at a lower temperature may give off a great amount of heat.

Generally speaking, temperature is a measure of a body's ability to either give up its heat to, or to absorb heat from, other bodies. Drop an ice cube into hot

Left: Boiling water in a large, full teakettle gives off more heat than water the same temperature in a small tea cup. Right: Water molecules frozen into ice are in constant movement; this movement increases as the ice heats, or melts.

tea, and two things happen: The tea gives up heat to the ice, and the ice absorbs heat from the tea. There is a temperature difference between the two bodies. After the ice melts and the tea cools, the mixture reaches thermal equilibrium, which will be at an intermediate temperature. The word "temperature" actually has different specific definitions, depending on methods used for measuring it.

What is felt as heat or cold from a body, however, is actually the body's thermal energy: specifically, the kinetic energy of its constituent atoms and molecules resulting from their constant motion. Even when fixed in a crystal lattice of solid ice, water molecules are jiggling in place—and that movement grows faster and more furious if the ice is heated. Eventually the kinetic energy of the water

molecules is so high that it breaks the molecular bonds holding them in a crystal lattice; that's when the ice begins to melt. When the liquid water is heated, its molecules move ever faster until—at water's boiling point—they begin to escape as a gas, or steam, in the case of water.

Thermal energy, or heat felt, is related to a body's density (mass and volume), whereas temperature is not. The Sun's beautiful outer corona—the silvery halo seen during a total solar eclipse—has a much higher temperature (1,800,000°F, or 1,000,000°C) than the surface of the Sun (about 10,000°F, or 5,500°C). But the corona's density is so low, maybe only 1 atom per cubic centimeter—scarcely greater than that of interplanetary space—that if you could put your hand into it, it would not even feel warm. Perception

of heat depends not only on the kinetic energy of individual atoms, but also on the density of molecules hitting your skin.

EXPANDING AND CONTRACTING

Two other temperature-related concepts are vital to understanding weather and climate.

First, with few exceptions, solids and liquids will expand when heated, and contract when cooled, the amount of expansion and contraction varying with the substance. Such thermal expansion happens because the molecular bonds holding the constituent atoms and molecules behave rather like springs, and adding heat increases the average distance between them. Thermal expansion accounts for why the mercury rises in the column of an old-fashioned fever thermometer or weather thermometer; the gradations along the column are calibrated to temperatures in degrees corresponding to different amounts of expansion.

Second, air and other gases also expand when heated and contract when cooled; the hotter the gas, the faster its particles move. If the gas is confined, the pressure, or force per unit area inside the container will increase because more particles will be bouncing off the walls every second. If the gas is free, or unconfined, however, the fast-moving particles will tend to fly away from one another, and the density will decrease. That's why higher layers of the atmosphere are thinner (less dense) than lower layers.

The Lower and Middle Atmospheres

The bottom two layers of the atmosphere—the troposphere and the stratosphere—account for 99 percent of the atmosphere by mass, as well as 99 percent of its water vapor. Above them are three other atmospheric layers important to life on Earth. But most weather occurs in the bottom layer: the troposphere.

TROPOSPHERE

Closest to Earth is the troposphere. Beginning at Earth's surface, its height varies, ranging from just under 4 miles (6 km) at the poles to nearly 12 miles (20 km) at the Equator. The troposphere averages about 7 miles thick (11 km) at the mid-latitudes of 50 degrees north, at London, for example, and 50 degrees south, the latitude of Patagonia.

Comprising 80 percent of the mass of the atmosphere, the troposphere is the air we breathe. It is composed of 78 percent nitrogen, 21 percent oxygen, traces of other gases, dust, other particles, and nearly 99 percent of the atmosphere's water vapor.

The air closer to the ground is warmer because land absorbs sunlight and heats up faster than the air above it; updrafts then carry that heat up into the air largely by convection (vertical air currents). The troposphere is thus the home of most weather; the air's vertical circulation results in the formation of clouds, while its horizontal movements result in wind.

As any mountain-climber can attest, the troposphere gets thinner with height, meaning the gases at higher altitudes get less dense. It also gets dryer and colder. The temperature drops from about 59°F (15°C) at Earth's surface to -71°F (-57°C) at the top of the troposphere. The troposphere is considered to end where temperature stops dropping with height. This region, known as the tropopause, marks the transition to the next layer, the stratosphere. Together, the troposphere and tropopause are known as the lower atmosphere.

STRATOSPHERE

The stratosphere, accounting for about 19 percent of the mass of the atmosphere, starts just above the tropopause and extends to about 30 miles (50 km) above Earth.

In this layer, the air temperature remains relatively constant up to an altitude of about 15 miles (25 km), and then actually increases with height, from an average -71°F (-57°C) at the tropopause to higher than

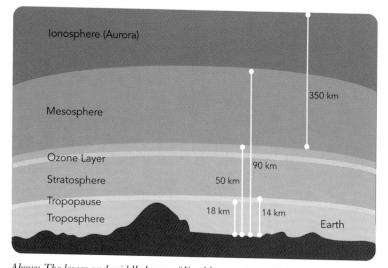

Above: The lower and middle layers of Earth's atmosphere. Top left: Most commercial airlines fly just below tropopause.

5°F (-15°C) at the stratopause. This increase in temperature tends to have a stabilizing effect on atmospheric conditions, making the stratosphere calm, so many jet aircraft fly in this layer.

Why is the stratosphere so hot? It is because molecules of oxygen (O_2) do two things. First, they scatter much of the solar energy—that is, incoming photons of solar ultraviolet radiation kick the oxygen molecules into physically moving faster, so their energy is converted to kinetic energy (higher temperatures). Second, the oxygen molecules also absorb much of the Sun's ultraviolet radiation, chemically changing to form molecules of ozone (O_3). Indeed, the stratosphere contains nearly 90 percent of the atmospheric ozone; that high-altitude ozone layer is essential for life on Earth, as it shields plants, animals, and people from much harmful solar UV radiation.

The stratosphere ends at the stratopause, where temperature once again starts dropping with increased height.

MESOSPHERE

The mesosphere, the middle layer of the atmospheric five-layer cake, extends from the stratopause to something over 50 miles (80 km) above the Earth—the nominal edge of space, according to one definition (see sidebar "Where Does Space Begin?").

The mesosphere is so rarefied that it accounts for only 0.1 percent of the mass of the atmosphere as a whole.

Even so, the mesosphere is still thick enough to slow meteors hurtling into the atmosphere, where they burn up, leaving fiery trains in the night sky. That protects the surface of Earth from being virtually sand-blasted by the tons of interplanetary grit that falls into the atmosphere each day. Moreover, there are so few oxygen molecules that the warming by absorption of solar UV also lessens. Thus, in the mesosphere, temperature falls with height, plunging from about 5°F (-15°C) at the stratopause to as low as -120°F (-85°C) at the mesopause, making the mesosphere the coldest layer of Earth's atmosphere.

The stratosphere and mesosphere, along with the stratopause and mesopause, are together known to scientists as the middle atmosphere.

Theodore von Kármán (1881–1963), the Hungarian aeronautical engineer and physicist who helped define the boundaries of outer space.

WHERE DOES SPACE BEGIN?

Earth's atmosphere does not end abruptly at a fixed altitude. So where outer space begins is subject to debate, and depends on what authorities you choose to believe.

Since the beginning of the space age in the late 1950s, U.S. authorities have defined the edge of space as a height of 50 miles (about 80 km) above mean sea level, about where the mesosphere ends.

This definition, however, is not commonly accepted internationally. The Fédération Aéronautique Internationale (FAI), the international air sports organization, chooses instead to define outer space as beginning at a height of 100 km (about 62 miles) above mean sea level. That height is called the Kármán line after Hungarian-born aeronautical pioneer Theodore von Kármán, who was seeking to calculate at what altitude a vehicle would have to travel faster than orbital velocity in order to achieve enough lift from the thin atmosphere to stay aloft. Although the answer he derived was not exactly 100 km, he proposed 100 km as a memorable round number for an altitude where Earth's atmosphere becomes negligible for aeronautical purposes.

Upper Atmosphere

Although they are sometimes considered part of outer space, the upper two layers of the atmosphere play important roles for Earth. The lower layer, the thermosphere, is where the northern and southern lights called the aurora flicker in eerie colors, and AM radio broadcasts are reflected back toward the ground at night. The uppermost layer—the exosphere—is essentially the same as the atmosphere that actually surrounds the so-called airless planets, such as Mercury and the Moon.

THERMOSPHERE

Right above the coldest layer of the atmosphere (the mesosphere) is the hottest layer: the thermosphere.

The thermosphere starts just above the mesopause at an altitude of about 50 miles (80 km), and extends up to perhaps 370 miles (600 km). The temperatures keep increasing with altitude until near the top of this layer, where they can reach as high as 3,600°F (2,000°C). The thermosphere is known as the upper atmosphere. Its

temperatures skyrocket because the few remaining molecules of oxygen absorb the most intense high-energy UV and X-ray radiation from the Sun.

The thermosphere is home to the ionosphere, a layer of electrically charged (ionized) gas particles that extends from about 35 to 190 miles (60 to 300 km) above the Earth's surface. The ionosphere was discovered in the early twentieth century during the early days of radio,

when it was important because it reflected radio waves and thus allowed certain radio transmissions to be heard thousands of miles away around the world (see sidebar "Listening to the Atmosphere").

It is also in the thermosphere that the aurora borealis and aurora australis, the northern and southern lights, flicker and dance. The aurora arises when charged particles (mostly electrons) from the Sun reach

Above: The northern lights, or aurora borealis, arise when charged solar particles enter the Earth's atmosphere. Top left: Artwork of the International Space Station orbiting Earth in the thermosphere.

Earth's magnetic field and are accelerated downward toward Earth's poles. Those charged particles collide violently with atoms and molecules in the thermosphere, energizing them and causing them to glow like the gases in a neon sign. The lurid colors of the aurora are produced by different atoms: the characteristic yellow-greens usually come from oxygen, while reds are often produced by nitrogen.

Many spacecraft—including the space shuttle, the International Space Station, and the Hubble Space Telescope—are in the thermosphere when they orbit Earth.

The thermosphere is separated from the last, outermost layer by the thermopause.

EXOSPHERE

The exosphere is the outermost layer of the atmosphere. Its height is ill-defined, depending not only on the atmospheric model and definition used, but also on solar activity and other physical characteristics. Some sources define it as extending from the thermopause to as high an altitude of 6,200 miles (10,000 km), about as high as could be said that an atmospheric molecule were in orbit around Earth. In the exosphere, atoms and molecules escape into space and Earth's atmospheric gases mix with the molecules of interplanetary gases and the charged particles from the Sun (the solar wind). Hydrogen and helium are the exosphere's prime components and are present only at very low densities.

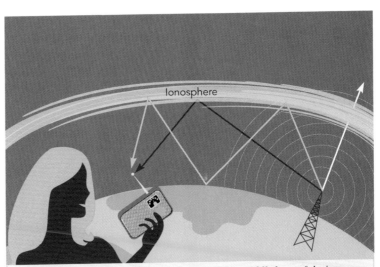

Radio signals are transmitted by bouncing off the middle layer of the ionosphere, called the E-layer, as discovered in 1901 by Guglielmo Marconi.

LISTENING TO THE ATMOSPHERE

The ionosphere, the layer of electrically charged gas particles in the outer atmosphere, exists because intense solar ultraviolet and X-ray radiation ionizes (breaks apart) atoms and molecules of Earth's upper atmosphere, giving them an electrical charge. The ionosphere has several layers, whose densities, altitudes, and electrical properties (affecting what radio frequencies they can reflect) change from day to night, from summer to winter, and with the intensity of solar activity (sunspot cycle).

The lowest layer is the D-layer 30 to 55 miles (50 to 90 km) high, which forms in the daytime and disappears at night. It absorbs AM radio signals, preventing them from bouncing off the higher layers in the daytime.

The middle layer is the E-layer, discovered by Guglielmo Marconi in one of his first radio experiments in 1901. One night, Marconi transmitted a signal between Europe and North America; he calculated that the signal had to have bounced off an electrically conducting layer about 62 miles (100 km) high. Actually, the E-layer ranges from about 55 miles (90 km) in the daytime to about 75 miles (120 km) at night, part of the reason that radio signals bouncing off it can travel farther at night. It is best at reflecting radio frequencies lower than 10 MHz. There's also an occasional patchy lower-altitude E_s layer ("S" for sporadic) that can form in the daytime, especially in summer, and can reflect higher radio frequencies (25 to 225 MHz)—very exciting to amateur radio operators, who are always trying to see where in the world they can communicate.

The upper layer of the ionosphere is the F-layer, ranging from 75 to 250 miles' (120 to 400 km) altitude. In the daytime, energetic rays from the Sun cause it to split into two layers (F_1 and F_2), which recombine at night after sunset.

Composition of the Atmosphere

Air is all around us: invisible, intangible, odorless—essential to all life on Earth. But air is actually a mixture of elements. By volume, assuming a dry atmosphere, the three main gases in the troposphere and stratosphere are nitrogen (78 percent), oxygen (21 percent), and argon (0.9 percent).

The remaining 0.1 percent of the dry atmosphere consists of trace constituents: carbon dioxide, ozone, methane, and various oxides of nitrogen, neon, and helium. Although percentages are small compared to the atmosphere's global volume, the first three are enormously important to life, weather, and climate, especially since many are concentrated in certain atmospheric layers.

Not included in these dry-atmosphere percentages is water vapor, which in the troposphere varies widely from 1 to 4 percent.

WATER VAPOR

Water vapor (H_2O), because of its properties and its abundance in the troposphere, is essential to meteorology and climate.

In the troposphere, it absorbs and reradiates thermal (infrared) energy from Earth's surface. Water vapor also plays a key role in the formation of aerosols in the stratosphere. Aerosols are tiny liquid droplets or solid particles that can remain suspended in the atmosphere a long time.

At very low temperatures, water vapor also forms special types of clouds known as polar stratospheric clouds over Antarctica in winter, which have been linked to the destructive hole in the ozone layer, with possible consequences for life on Earth (see chapter 12).

Water vapor consists of only 0.0004–0.0006 percent of the stratosphere—percentages so small that scientists prefer to specify measurements as 4 to 6 parts per million of volume (ppmv). Even at these exceptionally low concentrations in the stratosphere, water vapor plays a significant role in the energy budget of the atmosphere.

CARBON DIOXIDE

Carbon dioxide (CO_2) makes up a global average of 381 ppmv (0.0381 percent of the dry atmosphere), concentrated at the bottom of the troposphere. Whereas animals inhale oxygen and exhale carbon dioxide, in a life-giving symbiosis, photosynthesizing plants in bright sunlight—especially young trees in fast-growing forests—take up carbon dioxide and give off oxygen. The gas is also absorbed

Above: Polar stratospheric clouds, formed by water vapor at very low temperatures, can form over Antarctica in winter. Top left: Biomass burning to clear land for agriculture can result in dangerously high levels of ozone near ground level.

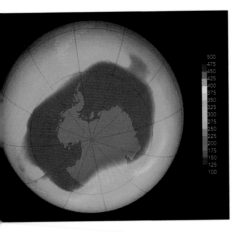

One of the largest ozone holes ever recorded, from data recorded by NASA's Aura satellite during the period of September 21–30 of 2006.

by the oceans in the cool polar regions. Carbon dioxide has an important effect on climate, by playing a role in regulating atmospheric temperature.

OZONE

That stratospheric ozone layer is responsible for life as we know it on Earth. It absorbs energetic solar ultraviolet rays that would harm many species if they penetrated to the ground. Its stratospheric concentration can reach as high as 10 ppmv (0.001 percent).

Nonetheless, ozone (O_3) itself is unhealthful to breathe, and should be avoided at ground level. It appears close to the ground as a result of human activities. In developing nations, ozone is released from biomass burning: the burning of jungle, savannah, and forests, primarily from humans clearing land for agriculture. In industrialized nations, it arises in photochemical smog from industrial pollution and vehicle exhaust reacting with sunlight. It may be a significant factor in climate change.

In the stratosphere, however, ozone is essential to life on Earth, because it blocks most harmful ultraviolet rays. For this reason, over the last several decades, meteorologists have been concerned about a large hole in the ozone layer that has persisted over Antarctica and high southern latitudes.

METHANE

On land, the main sources of methane (CH_4) are rice paddies and—as humorous as it may sound—the flatulence of cattle. In the ocean, a main source may be undersea cold seeps and eruptions.

Although methane has a global average of 1.7 ppmv (0.00017 percent) in the well-mixed lower atmosphere, it is more concentrated near its sources. Measurements of the chemical composition of bubbles in ice cores dug up in the Arctic and Antarctic show that methane concentrations remained constant at about 0.7 ppmv for thousands of years and has increased to 1.7 ppmv only relatively recently on geological time scales. Even at low concentrations, it may be a powerful factor in climate change.

AEROSOLS AND IONS

In addition to gas molecules in the atmosphere, there are also suspended solid or liquid particulates. Known as aerosols, they range from micrometers to millimeters in size. Natural aerosols include dust, sea salt, and even tiny droplets of sulfuric acid (same stuff as battery acid), this last resulting from chemical reactions from volcanic sulfur dioxide. Anthropogenic (man-made) aerosols are usually pollutants, such as sulfur dioxide from power plant emissions. Aerosols may have significant consequences on the climate because of ways they absorb and reflect solar energy, form nuclei around which raindrops and hailstones can form, and also provide surfaces for chemical reactions (such as those leading to the destruction of ozone in the stratosphere).

Ions are atoms or molecules with an electric charge as the result of gaining or losing electrons. Produced by different energetic phenomena, they are present at all altitudes in the atmosphere. In the troposphere, ions are produced by lightning, cosmic rays, and even by the decay of radioactive elements in Earth's crust. In the middle and upper atmosphere, high-speed charged particles (primarily protons) from the solar wind also ionize some chemical compounds.

Methane emissions from rice paddies are considered to be a factor in human-induced climate change.

13

A Variety of Perspectives

In common speech, people often use the terms "weather" and "climate" interchangeably, but these words describe very different phenomena. Weather is the state of the atmosphere here and now—today, tomorrow, next week. Climate is the characteristic pattern of weather for a certain region or for the entire world over long periods of time, ranging from several decades to 100,000 years.

The tools of meteorologists and climatologists are often the same, but also sometimes different or specialized.

Above: A small weather station on a Swiss mountain summit. Top left: A Doppler radar dome.

STUDYING WEATHER

For recording local weather, thousands of tiny stations—often a white louvered box a couple of feet on a side housing automated instruments—are scattered around the nation at local airports, astronomical observatories, even isolated mountain summits. They record high and low temperatures, rainfall, and sometimes wind speed and direction and barometric air pressure; they then transmit the readings to meteorologists.

In mountains where snowfall is important to skiing, or spring runoff is important to local water supplies, other instruments and experts measure inches of snow and rain. At airports and major weather stations, Doppler radar systems can detect both the movement of thunderstorms miles away, and the approach of local rain and strong winds. Permanent records are made of such local readings, which are helpful for understanding a region's climate and any changes in the climate over time. These readings also provide weather radio stations with fascinating facts about each day's record high and low temperatures, rain and snowfall records, and other data.

For short-term weather forecasts of a day or two, or long-term weather forecasts over

National Hurricane Center director Max Mayfield with a radar image of Hurricane Frances, September 2004.

a week and a half, meteorologists rely not just on the ongoing record of local reports, but on two other critical tools.

First, they study a big picture of developing weather systems photographed by weather satellites in geostationary orbit above the Equator. These satellites orbit Earth once in 24 hours, thereby appearing to remain essentially stationary above one spot on Earth. In addition to documenting developing structures of clouds by visible light, weather satellites also gather data at other useful wavelengths in the electromagnetic spectrum, such as infrared (heat) or ultraviolet radiation. For tracing the development, movement, and violence of thunderstorms, there is even an automated network that records the time and location of every lightning strike around the nation. Moreover, at the first signs of violent weather, a daring breed of meteorologists dubbed storm chasers risk their own lives to fly or drive exceptionally near or into thunderstorms and hurricanes, to get middle-altitude readings unattainable any other way.

Second, meteorologists rely on mathematical models of weather systems that have been developed over the decades, on high-speed computers. Indeed, the mathematical modeling and simulation of weather systems absolutely depends on high-performance computers capable of handling massive amounts of data and calculating predictions faster than real-life weather systems actually move.

STUDYING CLIMATE

Climate studies rely not only on historical weather records, but also on awareness of

developments in other disciplines. For example, after the April 1815 volcanic eruption of Mount Tambora in Indonesia, 1816 became known as "the year without a summer," with snow falling across the United States even in July. Similarly, for several years after the eruption of the Indonesian volcano Krakatau in 1883—the second most violent volcanic eruption of the nineteenth century, after Mt. Tambora—people even in Europe and the United States were treated to spectacular red and orange sunsets and other meteorological phenomena. Modern studies of those reports have given present-day atmospheric scientists insights into how volcanic ash and aerosols (such as microscopic droplets of sulfuric acid) were carried around the world by stratospheric winds. Moreover,

Scientists study tree rings for clues to past climate change. Wider rings correspond to years of higher rainfall.

astronomers' telescopic observations of a virtual absence of sunspots from 1645 to 1715 is correlated with the coldest 70 years of the Little Ice Age that gripped Europe and North America from the Renaissance into the middle of the nineteenth century.

Climatologists supplement written records by looking for evidence of past climate change in the geological record. They core for deep samples of ice in the ancient ice sheets of Greenland and Antarctica, which preserve records of volcanic eruptions and other climate changes going back millennia. They measure the concentration of carbon-14 (a radioactive isotope of carbon) in the tree rings of 2,000-year-old trees to quantify solar activity over time. They are also exploring astronomical evidence that changes in the tilt of Earth's rotational axis as well as in the shape of Earth's orbit around the Sun may have altered climate over cycles of 10,000 to 100,000 years.

Scene on the Ice by the Dutch painter Hendrick Avercamp (1585–1634). The cold period known as the Little Ice Age corresponded with a virtual absence of sunspots, and was depicted in many European paintings of the period.

Intertwined Futures

The atmosphere is not an isolated system. Air flows freely across all human borders, regardless of boundaries between cities, counties, states, and nations. It also flows around and over major natural obstacles, including the Pacific and Atlantic oceans, eventually encircling the globe. It makes its way to the poles and into the stratosphere, interacting with the oceans and land, with still-unforeseen consequences both for health and climate. In short, air pollution generated in one place can be inhaled not only by nearby humans, animals, and plants, but even half the world away; smog from cars in the Chinese capital of Beijing has been detected as far away as California.

One of the most important interactions of the atmosphere with the ocean and land is the so-called greenhouse effect (see chapter 5). Water vapor, carbon dioxide, and methane—although extremely small percentages by comparison to the entire atmospheric mass—are enormously important in trapping solar energy and keeping the net temperature of planet Earth above freezing. Without the greenhouse effect, the oceans would have long since frozen solid and life could not have evolved.

For millennia, up until about the middle of the nineteenth century, most changes in weather and climate were due to natural events: the spontaneous upwelling of warmer or colder waters in the oceans, volcanic eruptions, variations in the Sun's radiation or the Earth's rotation or orbital revolution, or even the impact of an asteroid on Earth, such as the event hundreds of millions of years ago that may have ended the Age of Dinosaurs.

Over the last century and a half, however, human activity, and its environmental impact, has begun dominating the state of the atmosphere. And that raises grave questions.

HUMAN IMPACT ON CLIMATE

Beginning with the widespread burning of coal during the Industrial Revolution in Europe and the United States, considerable amounts of carbon dioxide and greenhouse gases (gases that contribute to atmospheric warming due to the greenhouse effect) have been released into the air—an amount that has continued to increase with the popularity of the automobile.

Although clean-air regulations enacted in the 1970s have helped significantly in

Above: A rickshaw in Calcutta, India. Top left: Heavy traffic in Taiwan. Cars are rapidly replacing human- and animal-powered transportation in developing countries; the resulting carbon emissions are a major factor in climate change.

curbing pollution from factories and cars, the population of the United States continued to grow and ever more Americans took to automobiles. In the nineteen-eighties and nineties, they were joined by growing populations in developing nations who also wanted to enjoy an American-style material life, and so set aside their oxcarts and bicycles in favor of a car. In consequence, in the twentieth century alone, more petroleum and coal was consumed than in all previous centuries combined.

The pace of such consumption, instead of tailing off, is accelerating as nations industrialize and as governments are reluctant to enact legislation. Not only does industrialization give rise to thousands of point sources of air pollution, but agricultural or urban development also entails the clear-cutting of vast tracts of forests (such as in the Amazon jungle)—destroying trees that otherwise would convert at least some of the carbon dioxide into oxygen through photosynthesis, cleansing air for the entire planet.

GLOBAL WARNING

In the 1970s and '80s, scientific evidence was still not definitive. But by the early twenty-first century, evidence was mounting that human-released greenhouse gases were having an impact on rising global temperatures, with the potential for consequences for weather and climate.

Here are a few data points: NASA reported in 2006 that the five hottest years on record worldwide occurred between 1995 and 2005. Atmospheric carbon dioxide has increased by 30 percent since the Industrial Revolution. Average global temperature is now rising at a rate of one to three degrees Celsius per century. Because temperature and carbon dioxide levels have matched each other in the past, adding more carbon dioxide to the atmosphere will likely lead to warming.

Above: A section of the Greenland ice sheet. Global warming is melting these ancient glaciers, resulting in rising sea levels worldwide. Bottom: An adult polar bear photographed in the Svalbard Archipelago in the Arctic Ocean. This iconic species is imperiled by reduced hunting grounds, as global warming melts polar ice.

Glaciers are retreating. Cubic miles of ancient ice sheets in Greenland and Antarctica have calved off into the ocean. Worldwide, the overall temperature of the ocean has increased by 1.5°F (1°C), which represents a staggering amount of heat energy, considering the volume and mass of so much water. Because the heat of ocean water is one factor driving the intensity of hurricanes, many wondered whether hurricanes could become more powerful in the future.

CHAPTER 2

METEOROLOGY'S ORIGINS

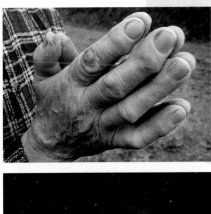

Left: A woodcut showing the Renaissance concept of cosmology, with the dome of heaven seen as a finite boundary. The universe beyond was viewed as the realm of the divine. The woodcut is sometimes attributed to nineteenth-century French astronomer Camille Flammarion, while other scholars accept it as an original artwork from the sixteenth century. Top: Weather is crucial to farmers' livelihood; before accurate weather forecasting, many consulted signs such as aching joints to predict the weather. Bottom: Comets were puzzling and awe-inspiring to ancient peoples; some viewed them as 'exhalations' from the Earth.

People in ages past turned to any sign or signal they could find to predict the weather. For example, they might notice that their corns or joints started aching with the approach of unsettled or stormy weather. Since one guess seemed as good as another, by the seventeenth century, farmer's almanacs often printed predictions of weather a whole year in advance. Pre-Galilean astronomy had been dominated by Aristotle's concepts of the immutable perfection of the heavens; as late as the seventeenth century some wondered whether comets or other astronomical phenomena were actually meteorological "exhalations" from the ever-changing and corruptible Earth.

The Enlightenment of the seventeenth and eighteenth centuries was a remarkable period of scientific discovery. Great advances were made in meteorology, seeing the invention of the thermometer, and concepts of temperature; the barometer, and concepts of air pressure and causes of winds; and the hygrometer, and concepts of humidity. Such momentous scientific discoveries and invention continued apace throughout the twentieth century.

Quantifying Heat and Cold

Over the centuries, an ingenious variety of thermometers were devised by a number of prominent scientists, among them astronomers Galileo and Olaus Roemer. But many had practical challenges, the usual one being that they were difficult to calibrate. Even though individual thermometers had become quite clever and advanced in design, scientists still needed a standard thermometric scale that would allow people at different locations to compare temperatures.

FAHRENHEIT SCALE

The most widely used temperature scale in the United States today dates back to the

Above: Galileo Galilei (1564–1642), the Italian physicist and astronomer, devised an early thermometer. Top left: The boiling point of water is a constant temperature, useful in quantifying heat.

German physicist Daniel Gabriel Fahrenheit (1686–1736), who did most of his work in Amsterdam and The Hague. A glassblower who came to specialize in fashioning precise scientific instruments, he devised several meteorological instruments, including altimeters, barometers, and thermometers. But his major contribution was his Fahrenheit thermometric scale, which—in the variant most widely accepted after his death—set water's boiling point at 212° and its freezing point at 32°. The Fahrenheit scale was also widely used in Europe until supplanted by the Celsius scale in the nineteenth century.

CELSIUS SCALE

The temperature scale now used in most sciences dates back to Swedish astronomer Anders Celsius (1701–44), who proposed it just two years before his untimely death from tuberculosis. He proposed that what he called the two "constant degrees" on a thermometer—the freezing point and boiling point of water—should be called 0° and 100° (although Celsius himself suggested that the scale be upside down from what it is today, with 0° being the boiling point). In the United States, the scale has also been called the

Centigrade scale, named for its 100 divisions between the two constant degrees. Celsius is now the officially recognized name for this temperature scale, and it is the standard in Europe and most of the world outside the United States.

Galileo's thermometer was a sealed glass tube filled with water and several blown-glass bubbles filled with colored liquid. The bubbles respond to the changing density of water as it heats or cools.

Thermodynamic Equilibrium
(Zeroth Law)

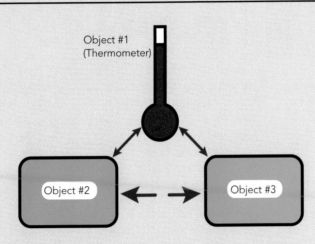

Object #1
(Thermometer)

Object #2

Object #3

The zeroth law of thermodynamics. When two objects are in thermal balance with a third object, they must be in balance with one another. This law is the underlying principle in how thermometers work.

THE ZEROTH LAW

If Agnes knows Carter, and if Beatrice also knows Carter, do Agnes and Beatrice know each other? Not necessarily. If piece of iron A is attracted to a magnet C, and another chunk of iron B is also attracted to magnet C, do chunks of iron A and B attract each other? Definitely not.

So why should thermometers work? Underlying any discussion of temperature in meteorology are several axioms or laws fundamental to all physics. Those assumptions are called the laws of thermodynamics: laws dealing with the relationship of heat to other forms of energy, as well as to such quantities as temperature, pressure, and volume. The laws of thermodynamics hold as true for hurricanes as they do for steam engines, and are used by meteorologists in their mathematical models of weather and climate. Officially there are three laws of thermodynamics. But underlying them all is a fourth—often called the "zeroth law of thermodynamics."

The zeroth law of thermodynamics: If bodies A and B are in thermal equilibrium with a third body C, then A and B are in thermal equilibrium with each other. Fundamentally, the zeroth law says that there is a useful quantity called temperature, and that thermometers work.

Heat energy naturally transfers only from hotter bodies to cooler bodies by conduction, convection, or radiation (see chapter 3). If there is any temperature difference between bodies A and C, then eventually both bodies end up at the same intermediate temperature—that is, they reach thermal equilibrium, with no temperature difference between them.

Now, say one of those bodies (C) is a thermometer, which is much smaller and less massive than the body A. Then the reading of thermometer C indicates the temperature of body A—which is why a fever thermometer can indicate the temperature of a sick child (because the thermometer is so small compared to the child, it scarcely removes any heat from the child's body in coming up to the child's temperature).

Now, here's the crux: If thermometer C then gets the *same* reading for some other body B, that means that bodies A and B—even if widely separated in space or time—are at the same temperature.

Although this concept is taken for granted today, it is not obvious from other walks of life. Nor could it be known before the advent of standardized temperature scales in the eighteenth century, or the discovery of the laws of thermodynamics in the nineteenth century.

Measuring Air's Pressure

Even though we live at the bottom of an atmosphere hundreds of miles deep, we're seldom conscious of the fact that air has mass, because we move so easily through it. But it does. And it follows that if the air overhead has mass, it also weighs down on everything with some pressure. Normally, we don't feel this air pressure because it's nearly equal from all directions. But when hurricane winds snap tree limbs or unroof houses, we witness the enormous force that air can have.

TORRICELLI AND THE BAROMETER

The first person to measure atmospheric pressure was Evangelista Torricelli (1608–47), a brilliant and short-lived Italian mathematician and physicist who worked and corresponded with Galileo, even briefly living in his home. In 1643, while consulting with well-diggers on why they couldn't siphon water up from wells deeper than about 33 feet around sea level, Torricelli suspected that atmospheric pressure was the culprit. A siphon works like a drinking straw: If you insert a tube into a pool of water and partially evacuate the tube (whether by the action of lips and cheeks, or with a pump), water is pushed up the tube by the force of atmospheric pressure on the surrounding liquid.

If he were right, Torricelli reasoned, then at sea level the liquid metal mercury—which is 13.6 times as dense of water—would be forced up only one-13.6th as high in an evacuated tube as would water. So he took

A sculpture of Evangelista Toricelli (1608–47), who devised the first mercury barometer.

a glass tube about a yard long, sealed one end, and filled it with mercury. Blocking the open end with his finger, he upended the tube into a dish of mercury, and then removed his finger. Would all the mercury rush out into the bowl of water? Torricelli found that only a little of the mercury flowed out of the bottom of the tube, leaving a column of mercury about 30 inches above the top of the mercury in the bowl. Indeed, the pressure of the air pushing down on the surface of the mercury in the open bowl stopped the mercury in the tube from flowing down and out.

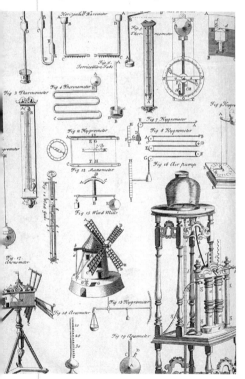

Above: A 1728 illustration shows various pneumatic devices and diagrams, all based on the principle that air has pressure. Top left: An aneroid barometer uses a flexible metal chamber and a spring to measure atmospheric pressure.

Torricelli thus proved that atmospheric pressure at sea level just counterbalances a column of mercury approximately 30 inches, or 76 centimeters (760 millimeters) high. In fact, we now know that high atmospheric pressure (which ushers in sunny and clear days) pushes down more on the mercury in the reservoir, forcing more mercury up the column; low pressure (which ushers in rainy days) pushes down less, so the level of mercury in the column drops.

PROOF BY PASCAL

Torricelli's belief in the pressure of the atmosphere was confirmed by his contemporary, French mathematician and physicist Blaise Pascal (1623–62). Pascal repeated Torricelli's experiment with many other liquids. He also hypothesized that if the height of the mercury column were sustained by atmospheric pressure, then the height should be less at higher altitudes, where atmospheric pressure is lower. So Pascal arranged to have a Torricellian apparatus carried to the top of a 3,000-foot mountain in central France. Indeed, there the mercury column was found to be about three inches lower at the summit than at the mountain's base.

Today, the Torricellian apparatus is called a mercury barometer (the prefix *baro-* or *bar-* coming from a Greek word meaning weight), although there are also many designs of barometers that do not use mercury or any other liquid.

Street sweepers in Manila, Philippines, work against the strong winds created by the low pressure of Typhoon Imbudo in 2003. The lowest atmospheric pressure on record was during a 1979 typhoon in this region.

PRESSURE BY ANY OTHER NAME

Both Torricelli and Pascal have units of pressure named for them—the torr and the pascal. To complicate matters, other units widely used in meteorology are inches or millimeters of mercury (inHg or mmHg, where Hg is the chemical symbol for mercury), and the bar—or, more usually, the millibar. They all start at the same reference point: atmospheric pressure at sea level. But after these units of pressure were originally defined, there were later calibration adjustments that ended up giving them odd quantities. Their equivalents are:

1 atmosphere = 1 torr = 760 millimeters of mercury (mmHg) = 29.9 inches of mercury (inHg) = 1.01325 bar = 1013.25 millibars (mbar) = 101,325 pascals (Pa) = 101.325 kilopascals (kPa) = 14.7 (14.695) pounds-force per square inch

The pascal is accepted by most nations, which use the International System (SI) of metric units; in the United States (often a hold-out about metric scientific measurements), however, the millibar and inches of mercury are still widely used and heard in radio and television weather reports.

Differences in weather and climate on Earth bring wide local variations in atmospheric pressure. The highest known atmospheric pressure, 108.6 kPa (1086 mbar or 32.06 inHg), was recorded on December 19, 2001 in Tosontsengel, Mongolia. Outside of tornado conditions, the lowest sea level atmospheric pressure, 87.0 kPa (870 mbar or 25.69 inHg), was recorded during Typhoon Tip in the western Pacific Ocean on October 12, 1979.

Catching the Wind

Air pressure exists even on a day that's dead calm. Wind is the horizontal movement of air. It exerts pressure, too, which can be measured using a pressure anemometer (from the Greek *anemos,* meaning wind). The wind speeds reported on radio and television weather forecasts, however, is wind speed in miles or kilometers per hour, measured with a velocity anemometer.

Another component for understanding a weather system or predicting the weather in a certain locale is wind direction. Wind direction is specified as the direction the wind is coming *from,* so a north wind is coming *from* the north. Most anemometers also have a vane or other device that freely rotates in any breeze to indicate the wind direction.

EARLY ANEMOMETERS
Although many civilizations have had devices for measuring wind, the first one usually cited is a swinging-plate anemometer invented around 1450 by Italian architect Leone Battista Alberti (1404–72). This mechanism was very much like the pressure-plate anemometer independently reinvented two centuries later by British mathematician and physicist Robert Hooke (1635–1702). In both instruments, a vane rotated to point

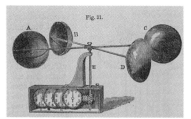

Above: A rotating cup anemometer, devised in 1846 by Irish astronomer Thomas Romney Robinson. Top left: A simple weathervane indicates wind direction.

Leone Battista Alberti (1404–72) is credited with inventing the first anemometer, a device for measuring wind.

in the wind direction, orienting a vertical disk or rectangular plate perpendicular to the wind; the angle by which the plate is displaced in the wind indicated the wind speed, rather like the angle of a sheet hanging on a clothesline on a breezy day gives a rough idea of the wind speed: The steeper the angle, the higher the wind speed.

The familiar rotating-cup anemometer seen atop many small weather stations and airports came two centuries later, in 1846, a product of the fertile imagination of Irish astronomer Thomas Romney Robinson (1792–1882), third director of the Armagh Observatory in Dublin. Using the principle of the windmill, but using the gearing to count rotations rather than to grind grain, Robinson devised a rotating-cup anemometer with four hemispherical cups, each mounted on the end of a horizontal rod. The rods are fixed to a rotating spindle; the speed of the rotation relates the speed of the wind. There are now three-cup versions as well. Sometimes the rotating cup anemometer is called a Friez anemometer after the founder of a commercial company that began manufacturing them around the turn of the twentieth century.

CLOCKING WINDS

Once you can measure wind speeds, how can those speeds be categorized? When does a tropical storm, for example, transition into becoming a full-fledged hurricane?

The first person to take a major systematic approach to this question was Irish hydrographer Sir Francis Beaufort (1774–1857), a captain in the British Royal Navy, who created what is now called the Beaufort scale for wind force. The Beaufort Sea above the Arctic Circle is also named for him. Beaufort ran away to sea at age 14. A year later, after being shipwrecked and nearly starving to death as the result of an inaccurate sea chart, he became obsessed with the importance of developing trustworthy charts for others similarly risking their lives. After mapping and charting many lands around the world and directing some major exploring expeditions, in 1829, he became Hydrographer for the British Admiralty, and spent the next 25 years converting what had been a minor chart repository into a superb surveying and charting institution.

A fully rigged historic schooner under sail. Sir Francis Beaufort (1774–1857) based his scale of wind levels on his observations of Royal Navy man-of-war ships.

Among other tasks aboard a ship at sea, naval officers recorded weather observations. But early in his career, Beaufort noticed that one man's "calm winds" might be another's "stiff breeze," so he recognized the need for a standardized scale. In 1805 he came up with a scale from Force 0 to Force 12, which was based on the effects of wind on the rigging of a Navy man-of-war; in the 1830s the scale became standard for Royal Navy ship's logs.

In the 1850s, the Beaufort scale was revised for non-naval use, with force numbers made to correspond with the revolutions of rotating-cup anemometers. In the twentieth century, other aspects of the Beaufort scale were standardized to meet the needs of steam ships, land observations, and meteorologists. The scale also was extended to

include Force 13 through 17, for rating tornadoes and Pacific typhoons. Today's categorization of hurricane-force winds corresponds to Force 12 (Category 1) through 16 (Category 5) on the Beaufort scale.

Beaufort Wind Scale	
Force 0	Calm
Force 1	Light Air
Force 2	Light Breeze
Force 3	Gentle Breeze
Force 4	Moderate Breeze
Force 5	Fresh Breeze
Force 6	Strong Breeze
Force 7	Near Gale
Force 8	Gale
Force 9	Strong Gale
Force 10	Strom
Force 11	Violent Storm
Force 12	Hurricane

Beaufort's wind scale, still in use today, and expanded to include hurricane force winds.

The NOAA ship Delaware II *navigates high seas. Whitecaps indicate higher wind levels.*

Quantifying Humidity

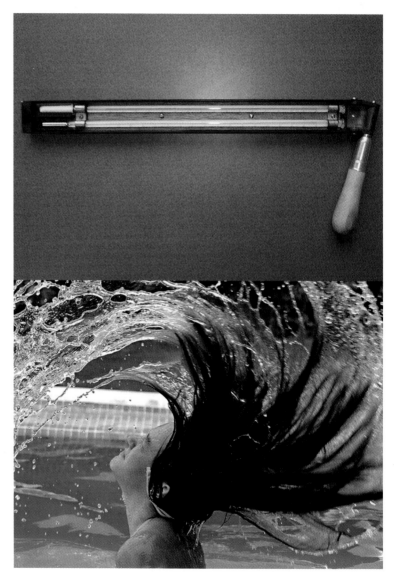

Although water vapor is a tiny fraction by mass of the atmosphere as a whole, virtually all of it is concentrated in the troposphere, especially at low altitudes. Indeed, in ground fogs, water vapor can account for 3 or 4 percent of a local air mass.

Water vapor is an important greenhouse gas, meaning that it helps air absorb heat radiated from the earth, trapping that heat and warming the lower atmosphere. Anyone can experience its effects: cloudy nights are generally warmer than clear nights, and humid days are hotter than dry days at the same temperature. Moreover, many weather processes are driven by, or result in, heat that is released or absorbed when water changes state—when it vaporizes, condenses, or freezes.

OF HAIR AND HYGROMETERS

Although Alberti, Hooke, Nicholas of Cusa, Leonardo da Vinci, and others proposed or experimented with ways of measuring dampness of the air, the person generally given credit for inventing a useful scientific hygrometer was Swiss geologist, botanist, and meteorologist Horace Bénédict de Saussure (1740–99). He observed that human hair, like some other

Top: A sling psychrometer, invented in the eighteenth century to measure humidity in the air. Bottom: Hair lengthens when it is wet, an observation that helped Horace Bénédict de Saussure (1740–99) invent a hair hygrometer.
Top left: Fog and mist are visible indicators of atmospheric humidity.

materials, is hygroscopic, or moisture-absorbing. More usefully, hair lengthens when it absorbs moisture and shortens when it dries. Saussure invented a hair hygrometer that indicated a humidity reading depending on how much a degreased hair lengthened or shortened. Hair hygrometers still find some uses today (and simple versions can be made at home).

In the nineteenth century, meteorologists came to prefer using what's now called a sling psychrometer; the prefix *psychro-* is derived from a Greek word meaning cold. The psychrometer is sometimes still called a wet- and dry-bulb thermometer, whose eighteenth-century inventor is not known.

As anyone who has emerged from a swimming pool can attest, evaporating water cools the body; the lower the humidity, the faster the evaporation, and the greater the cooling. A psychrometer takes advantage of that fact: Two identical mercury thermometers are mounted on a wooden paddle. The bulb of one thermometer is wrapped in an absorbent material that is saturated with distilled water. In the sling psychrometer, a person takes hold of a chain and handle

Dry, cracked earth can result from low rainfall, coupled with low humidity.

at one end of the wooden paddle and whirls the apparatus through the air. Because of evaporation, the thermometer with the wet bulb may be cooled dramatically compared to the one with the dry bulb. The temperature difference between

the two can be converted to relative humidity at that ambient temperature. A great difference, indicating faster evaporation, corresponds to lower relative humidity, while a minimal difference means the air is nearly saturated with moisture.

Air in the desert, while often quite hot, has a low level of relative humidity. Compare the dryness of desert air with humid regions of similar average temperatures, such as tropical rain forests.

HUMIDITY 101

Every weather report or forecast on radio or television gives numbers for "relative humidity," but what is that, exactly? Humidity relative to what? And is there an "absolute humidity" as well?

Absolute humidity is a measure of the actual amount of water vapor in a given volume of air, usually in terms of grams per cubic meter. But by itself, this figure is not very helpful to know because the absolute amount of moisture needed to saturate air varies significantly with the temperature. Warm air can hold a lot more water vapor than cold air. That's why air gets so dry whenever winter temperatures plunge below freezing. But not all warm air holds all the moisture it theoretically could: Think of desert air, for example, which is characteristically hot and dry.

Relative humidity is a percentage that represents the amount of water vapor *actually* in the air at a given temperature, divided by the maximum amount of water vapor the air theoretically *could* hold at that same temperature if it were fully saturated. It's this percentage that's correlated with how comfortable or uncomfortable the humidity feels at various temperatures. Usually, people will feel comfortable when relative humidity at room temperature is roughly 50 to 70 percent. On "muggy" days, the relative humidity may be 80 to 90 percent. Whether a day is hot or cold, a relative humidity of only 5 to 10 percent also comes to feel uncomfortable, as prolonged exposure can cause mucous membranes and even skin to dry and crack.

Mapping and Computing the Weather

By 1900, there were accurate standardized thermometers for measuring air temperature, barometers for measuring air pressure, anemometers for measuring wind speed and direction, and hygrometers for measuring humidity. Moreover, in the late nineteenth century, major nations had reached international agreements on calibrating instruments and procedures for taking readings so that information could be compared and shared.

All individual measurements, however, were highly local. Could they say anything about weather as a whole, changes in weather

to come, or overall patterns of weather and climate?

It was the telegraph that helped meteorologists perceive that weather seemed to have large patterns and movements. And it was early calculating machines that inspired some meteorologists to see if it were possible to predict those patterns.

TELEGRAPHY AND WEATHER

On May 24, 1844, Samuel F. B. Morse tapped out the question "What hath God wrought?" over the first telegraph line installed between Baltimore, Maryland, and Washington, D.C. Scarcely five years later, in 1849, Joseph Henry, founding secretary of the three-year-old Smithsonian Institution, persuaded telegraph companies to transmit regional weather reports free of charge. Aside from being a telegraphic pioneer himself, Henry had a long-standing interest in meteorology since his days as a physics professor at the Albany Academy in Albany, New York. In Albany, Henry had compiled reports of statewide meteorological observations for the University of the State of New York. Moreover, knowing that storms in the United States usually moved from west to

Top: Joseph Henry, the founding secretary of the Smithsonian Institution, and a meteorological pioneer, photographed in 1879. Bottom: A 1930 photograph of a cooperative weather station at Granger, Utah.

Above: The ENIAC (Electronic Numerical Integrator and Computer), the first large-scale, digital, programmable computer. The advent of such computers paved the way for accurate weather forecasting. Top left: An 1888 photograph of the Army Signal Service weather station, Cape Mendocino, California.

east, he saw that the telegraph might be used to warn people of advancing storms.

At about the same time, Henry also began building a network of volunteer weather stations around the United States. By the late 1850s, there were some 500 of these stations from New Orleans to New York. From their daily telegraphic dispatches, Henry devised a large daily weather map to (in his words) "show at one view the meteorological condition of the atmosphere over the whole country." The map, hung for public display in the Smithsonian Institution castle where it became a tourist attraction, was dotted with moveable colored disks showing local conditions (white for fair, blue for snow, black for rain, brown for clouds), along with arrows showing the direction of prevailing winds. Henry also shared the telegraph dispatches with the *Washington Evening Star,* which in May 1857 began publishing daily weather conditions at nearly 20 cities—the first popular newspaper weather page.

The Smithsonian's work was suspended by the Civil War. In 1870, Congress passed the responsibility for storm and weather predictions into the hands of the U.S. Army Signal Service; four years later, the military agency took over the volunteer observer system as well. In 1891, the weather functions of the Signal Service were transferred to the newly established U.S. Weather Bureau, which later became the National Weather Service.

A meteorologist in 1965 working at an IBM 7090 electronic computer in the Joint Numerical Weather Prediction unit. These computers processed weather data for forecasts, analysis, and research.

COMPUTING THE CLIMATE

In 1903, Vilhelm Bjerknes (1862–1951), a Norwegian physicist and meteorologist, began advocating a computational approach to weather forecasting. He believed it was possible to combine a full range of observation with a full range of newly developing theory to predict the weather. Shortly after World War I, Bjerknes and other meteorologists introduced the concepts of air masses, cold and warm fronts, and other concepts useful in forecasting.

The first person to make a full trial of Bjerknes's approach was English mathematical physicist Lewis Fry Richardson (1881–1950). After World War I, Richardson devised an algorthimic scheme for predicting the weather, laying the groundwork for much of what today is called numerical analysis. There was just one major problem:

calculating machines were so slow as to be useless—indeed, when Richardson tested the scheme, it took him six weeks to calculate a six-hour advance in the weather. His experience convinced meteorologists that a computational approach to weather prediction was completely impractical.

In World War II, meteorology came to be seen as having great military value. By this time, the new punched-card computers were fast enough to start to be a practical tool in forecasting. Indeed, Hungarian-born mathematician John von Neumann (1903–57) demonstrated at the Institute for Advanced Study in Princeton that powerful electronic computers could use physics-based algorithms to predict large-scale atmospheric motions as accurately as human forecasters. By the 1960s, computers became standard tools in meteorology.

A View from Above

The advent of the first weather satellites in the 1960s gave a huge boost to meteorology and climatology. For the first time, scientists could actually see weather systems in their entirety, which improved both practical weather forecasting and theoretical understanding of atmospheric dynamics.

Above: Two infrared images show the approach of Hurricane Andrew, which struck southern Florida on August 23, 1992. Top left: Artwork of the polar-orbiting environmental satellite NOAA-18, launched in 2005.

WEATHER FROM SATELLITES

The first successful weather satellite launched by the National Aeronautics and Space Administration (NASA) was TIROS-1 in 1960. It functioned for 78 days, and paved the way for the Nimbus program that followed. Today weather satellites are operated by the National Oceanic and Atmospheric Administration (NOAA). Information and images are also exchanged with other nations that launch and operate their own weather satellites, including China, India, Japan, Russia, and many European countries. This international meteorological cooperation, along with observations from other environmental-monitoring satellites allows a more complete picture of the weather to emerge than ever before.

Weather satellites see more than just cloud systems. They also can see and measure forest fires, water and air pollution, sand and dust storms, snow cover, ice floes, boundaries of ocean currents, flooding, clouds of volcanic ash, oil spills, changes in vegetation, and many other environmental factors that influence weather. They observe Earth not only in visible light, but also at thermal infrared

Historic photograph of TIROS-1 (Television Infrared Observation Satellite), the first successful weather satellite, launched in 1960.

(heat) and other wavelengths, enabling a scientist to determine cloud heights and types, to calculate land and surface water temperatures, and to locate ocean surface features.

DIFFERENT PURPOSES

Depending on their specific purpose and the types of observations needed, weather satellites may be launched into one of two types of orbits.

A sun-synchronous orbit is a circular polar (north-south)

orbit that carries the satellite over both poles at a relatively low altitude, typically 530 miles (850 km). Earth rotates beneath the satellite, so the satellite crosses over any given spot twice a day. This kind of polar orbit is useful for observing the entire surface of Earth up close at a consistent sun angle, to simplify the interpretation of shadows. Polar-orbiting satellites provide soundings of temperature and moisture through the entire depth of the atmosphere.

A geosynchronous orbit is a circular equatorial orbit at a high altitude of about 22,000 miles (35,000 km) that orbits Earth once in 24 hours; thus, the satellite effectively hangs almost directly over just one spot on Earth. It sees only a third of the planet, so three geostationary satellites would be necessary to monitor the entire planet. Geosynchronous orbit is useful for gaining a big synoptic, or comprehensive picture of major weather systems and their development.

SEEING THE UNSEEN

Weather satellites also can monitor phenomena that are widespread, or difficult or impossible to observe any other way. For example, weather satellites monitor El Niño and La Niña, periodic upwelling of warm or cold water in the Pacific off South America, linked with various changes in weather worldwide.

Weather satellites were what first drew attention to a hole in the stratospheric ozone layer over the Antarctic, which was later linked to deformities in animals in the southern

hemisphere and potential ill effects in humans. Scientists now agree that the destruction of the ozone layer has largely been caused by the human release of chlorofluorocarbons (CFCs) into the atmosphere—chemicals that in the 1980s were phased out of aerosol cans and the cooling coils of refrigerators.

In 1991, as Iraqi troops retreated from Kuwait, they lit over 700 oil wells on fire. The fires burned for several months, consuming millions of gallons of oil, and emitting plumes of smoke that stretched along the Persian Gulf coast into Saudi Arabia. Scientists were concerned with possible local climate effects of this smoke. Satellite images were taken by the Landsat-5

Chlorofluorocarbons (CFC's), phased out of aerosol cans in the 1980s, are believed to have contributed to holes in the ozone layer.

Thematic Mapper, and a C-131A aircraft equipped with an instrument called a Cloud Absorption Radiometer was sent into the smoke plume to measure scattered radiation. Initial results of the study showed locally damaging effects to air quality, and laid the groundwork for comparative studies of larger fires' effects on global climate.

Landsat images of the oil well fires in Kuwait lit by Iraqi troops in 1991, showing a plume of smoke extending into Saudi Arabia.

Networks of Data

Temperature, pressure, humidity, precipitation, and wind direction and speed are now often automatically recorded by untended weather stations and ocean buoys. Even more data are available with new types of instruments that measure and track other weather phenomena, helping meteorologists gain a more complete picture of weather on Earth.

Above: Historic photo of radiosonde weather balloons. Top left: A train whistle is higher-pitched when approaching than when receding, an example of the Doppler effect used in weather radar.

RADIOSONDES

Weather balloons—known as radiosondes—are helium- or hydrogen-filled balloons some 6 or 7 feet (about 2 m) across. They carry aloft instruments that measure wind speed and direction, humidity, temperature, and other characteristics of the atmosphere up into the tropopause. As the balloon ascends, a transmitter radios continuous measurements back to Earth throughout the atmosphere's depth, in finer detail than is possible even with today's polar-orbiting satellites.

Eventually, the radiosonde reaches a height where air pressure is so low that the balloon bursts, and the instrument package parachutes back to Earth. Instructions on the instrument package direct any finder to ship the package back to the weather station that launched it, as the latitude and longitude of its landing location also gives useful information about high-level winds.

Radiosondes are regularly released from certain meteorological research stations, such as in Antarctica. They may also be released for special purposes, such as just before a rocket launch as a final check for any turbulent winds over the launch site.

WEATHER RADAR

Radar, invented during World War II, helped win the war by allowing the Allies to detect the approach of German warplanes even at night or through clouds. In broad outline, weather radar uses a simple principle: an antenna transmits pulses of radio waves, and then waits to hear echoes returned.

Tuning the right frequency allows meteorologists to figure out the size and nature of the precipitation reflecting the energy, such as raindrops or hailstones. The direction of the echoes (determined by the

Radar technology is credited for helping the Allies win World War II, allowing them to detect the approach of German warplanes such as the Messerschmitt Bf 109E4 shown above.

rotating antenna) shows the compass direction of the precipitation, whereas the time elapsed between the pulse transmitted and the echo received tells the distance: The longer the time lapse, the greater the distance. Knowing that, the angular width of the echoed beam gives a clue about the size of the storm system, and the intensity of the echoes indicates the severity of the precipitation.

A Doppler radar can also detect whether pulse echoes are returning faster or slower than pulses sent. A higher frequency means that the precipitation is approaching, whereas a lower frequency means that it's receding, in the same way that the whistle of an approaching train is higher-pitched than the whistle of a train that is standing still or receding.

NUMERICAL WEATHER PREDICTION

After all the data are collected, then comes number-crunching. Scientists use computer simulations of the atmosphere called numerical weather-prediction models. Starting from the current set of data, and using the best available understanding of physics (thermodynamics) and fluid mechanics, they evolve the state of the atmosphere forward in time—kind of a meteorological version of computer-aging a child's face. Equations of how a fluid behaves—and air is a fluid—are so complex that supercomputers are needed. The computer model's output provides the basis for a weather

The HMS Beagle *in Tierra del Fuego, Argentina, painted by Conrad Martens during the* Beagle's *voyage of 1831–36. Robert FitzRoy, the ship's captain and meteorologist, also coined the term "weather forecast."*

FATHER OF FORECASTING

The term "weather forecast" was coined by Robert FitzRoy (1805–65), captain of the British Royal Navy ship HMS *Beagle,* including the *Beagle's* second voyage (1831–36) that carried young Charles Darwin. FitzRoy, himself a meteorological pioneer under Sir Francis Beaufort, devised the first portable weather station, and was charged with first using the Beaufort scale of wind force on the voyage that included Darwin.

Back on land, FitzRoy used the telegraph to gather daily observations from across England and developed *synoptic* weather charts allowing predictions to be made; indeed, FitzRoy coined the term *"weather forecast,"* publishing the first-ever daily weather forecasts in the *London Times* in 1860. The next year, he introduced a system of hoisting storm-warning cones at principal seaports when a gale was expected.

forecast, which can then be converted into graphics for public presentation on television.

Humans are best at interpreting the computer output especially for "nowcasting"—short-range weather forecasting from 6 to 24 hours ahead. Humans also can use knowledge of local effects that may not be programmed into a model. Even with the great advances in technology that have given us ever more accurate

weather forecasts, a person in the eastern U.S. can still feel the wind pick up, see the sky turn mustard yellow, and know a thunderstorm is imminent, while an observer in San Francisco can feel a dampness indicating the afternoon fog is about to roll in. So while we've come a long way from forecasting the weather by paying attention to our achy joints, the five human senses can still tell us much about weather on the way.

CHAPTER 3

OUR CONNECTION TO THE COSMOS

Left: Earth from space, seen in a true-color NASA image created by combining data from two satellites. Top: A view of galaxies far beyond our own Milky Way; the composite image comes from Two-Micron All-Sky Survey. Our own bluish-tinted galaxy lies toward the upper middle of the image. Bottom: Maple trees in Virginia turn brilliant colors in the fall. The seasons result from the tilt of Earth's axis relative to its orbit around the Sun.

Earth is first and foremost a planet. Although the globe's weather varies widely, much weather phenomena girdle the entire planet. Global weather systems are powered by radiation pouring into the atmosphere from the Sun. Had Earth formed closer to or farther from the Sun, or had the Sun been a hotter or cooler star, our planet might have been as uninhabitable as Venus or Mars.

Earth's atmosphere acts as an insulating blanket, blocking harmful high-energy ultraviolet and X rays while admitting visible light and infrared radiation (heat), trapping heat and distributing it around the globe. This insulating effect of the atmosphere protects us from brutal day-night swings of hot and cold, as on the airless Moon. Some weather phenomena result from Earth's unique astronomical characteristics: Jet streams and prevailing winds arise from its daily axial rotation, and seasons arise from the tilt of the Earth as it orbits the Sun.

Over millennia, long-term changes in Earth's mean temperature and climate may arise from changes to our astronomical position. Scientists believe that at least once, an asteroid slammed into Earth, triggering the catastrophic climate change that rendered most dinosaurs extinct. Both large and small events on Earth are indicators of how directly our planet is affected by its position in the cosmos.

The Solar Connection

Earth both receives energy from the Sun, and gives off energy into space. The balance that results from this give-and-take can be thought of as Earth's energy budget, analogous to a household budget: Salaries (solar energy) come in, and expenses (Earth's radiation) go out. Just as a steady source of income is essential to most households, a steady source of solar energy is essential to a stable climate on Earth. But just how stable is the Sun?

Astronomically speaking, the Sun is very stable, a middle-aged yellow dwarf star. Like the bowl of porridge in "Goldilocks and the Three Bears," among stars

This well-known 1851 painting, George Washington Crossing the Delaware *by American artist Emanuel Gottlieb Leutze, is historically accurate as a record of past climate change. The year 1776, depicted in the painting, falls in the period known as the Little Ice Age, when rivers froze solid. The Delaware rarely freezes today.*

the Sun is neither the hottest nor the coolest. It is neither the biggest nor smallest, neither the oldest nor youngest, and neither the brightest nor dimmest—all lucky for Earth. Had the Sun been small and dim and cool, Earth might have been too dark and frigid for human life. Conversely, had the Sun been big and bright and hot, it would have burned through all its nuclear fuel in just a few hundred million years. Instead, it is only halfway through its life cycle at an estimated 4.5 billion years—plenty of time for Earth to have cooled and for plants, animals, and humans to have evolved.

But stable doesn't mean changeless. The Sun—a thermonuclear furnace fusing hydrogen into helium—is a seething mass of hot gases, its surface ever roiling and erupting and sending out masses of charged particles that intercept Earth. Over decades and centuries, the Sun has undergone certain astronomical changes that are still not understood, but that have had dramatic effects on Earth's weather and climate.

SUNSPOT CYCLES

The Sun's best-known variation is the sunspot cycle of roughly 11 years. At what is known as solar minimum, whole days might go by with the disk of the Sun looking almost blank. At solar maximum, however, the Sun's face might be peppered with dozens of dark spots, some so huge as to be visible to the

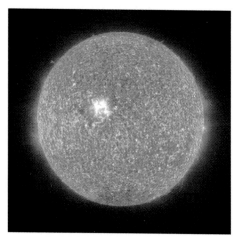

Above: Sunspot activity on the Sun occurs in a cycle of roughly 11 years, affecting Earth's climate. Top left: This ultraviolet image shows arcs of gas flowing around sunspots.

unaided eye. Today astronomers know that sunspots are colossal magnetic storms, hurricanes of electrified gas so enormous they could swallow several Earths; some sunspots are more than 30,000 miles (50,000 km) across.

The sunspot cycle affects Earth. In the years of solar maximum, the nighttime skies dance with ghostly green and red auroras (northern and southern lights), sometimes visible not just at higher latitudes but all the way to the Equator. These auroras are the result of highly charged particles from the Sun meeting the Earth's magnetosphere, where they collide with gases, causing them to glow. During the most dramatic auroral displays, radio, television, and satellite communications can be disrupted.

During sunspot maximum, the Sun actually glows about 0.1 percent brighter than at sunspot minimum. While this is not much, it is enough to affect Earth's climate. The same is true of sunspot minimums, when the Sun's brightness is diminished, an effect that became especially clear between 1645 and 1715, when sunspots and auroras all but disappeared.

Solar physicists call this 70-year period of solar inactivity and reduced solar brightness the Maunder minimum. The Maunder minimum corresponded to the most intense period of cold during the Little Ice Age, from the Renaissance through the mid-nineteenth century, during which rivers in Europe and the United States

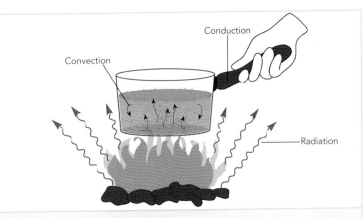

Diagram shows three different kinds of heat transfer: radiation from the fire, convection within the water, and conduction from the hot pot to its handle.

RADIATION, CONVECTION, AND CONDUCTION

Heat gets transferred into and out of the atmosphere in three ways: radiation, convection, and conduction. All are major factors in weather.

Radiation requires no medium; it is how solar electromagnetic energy—ranging from X rays to radio waves—is transferred across the vacuum of space to Earth. Examples of radiant energy on Earth are the light and heat from an incandescent light bulb; although the atmosphere is all around, the light bulb would also give off its heat and light in a vacuum. Dark-colored objects, which physicists call "black bodies," such as asphalt or charcoal absorb radiant heat and warm up faster than light-colored objects; they also reradiate heat and cool down faster.

Unlike radiation, conduction and convection both require a medium. Conduction is the transfer of heat within a material, or from one material to another through physical contact. Metals are excellent conductors, as many cooks have discovered after leaving a silver or steel spoon in a pot on a stove; the spoon's handle becomes too hot to touch. Air is a poor conductor. Water and rocks fall somewhere between.

Convection is the transfer of heat by the movement of mass, as occurs in boiling water. For example, radiant energy from the Sun warms rocks; the rocks heat the air directly above them, creating a warm bubble of air. Because warm air is less dense than cool air, the bubble of warm air rises, creating an air current that draws in cooler air over the rocks, and the process continues. Convection currents account for many weather patterns in the troposphere, from local winds to giant circulation cycles.

that didn't normally freeze became frozen solid, snowfields remained year-round at low latitudes, and patterns of agriculture and disease were altered. Lesser sunspot minimums match up with other dips in global temperature. What causes the sunspot cycle—or what disrupts it—and how the sunspot cycle is related to the Sun's energy output are still open questions.

Daily-Rotation Mixing

Where the Sun is at the zenith, it heats Earth and its atmosphere most. If Earth did not rotate on its axis, and if its axis were not tilted with respect to the plane of its orbit, that zenith hot spot would be over the Equator. Thus, the Equator would become very hot, and the heated, less-dense air would rise into the upper atmosphere. Cooler, denser air would be drawn in, and become heated in turn. Two large, stable convection cells would form, one in the northern hemisphere and one in the southern. In each convection cell, warm air would rise at the Equator and move toward the poles, where it would sink and return to the Equator.

But Earth does rotate on its axis. Moreover, because Earth is a solid sphere, higher latitudes rotate more slowly than lower latitudes, introducing horizontal forces (see sidebar "The Coriolis Effect"). As a result, Earth is girdled with three major convection cells in each hemisphere. Two of them (the Hadley cell and the Polar cell) exist as a direct result of the Sun's warming the surface of Earth. The third (the Ferrel cell) is more complex.

HADLEY CELL

Weather at tropical and subtropical latitudes is dominated by a convection cell that indeed starts with warm, moist air rising from the Equator into the tropopause, and then carried toward the poles. But about a third of the way there, at latitudes of about 30 to 60 degrees north or south, the warm air cools and descends. Some of the descending air travels along Earth's surface toward the Equator, closing the loop of each Hadley cell and creating the trade winds.

Evidence of Hadley Cell circulation is readily visible on satellite photographs of Earth. A band of cumulus clouds near the Equator show where the warm, moist equatorial air is rising. Paralleling it both north and south are bands of clear skies at subtropical latitudes, showing where cooled, dry air is descending.

Indeed, all the great deserts of the world are at subtropical latitudes where dry air at the northern or southern edge of the Hadley cell is descending—the southwestern United States and the Sahara in the northern hemisphere, and the great Australian desert and Chile's arid Atacama Desert in the southern.

The equatorial zone has not just one Hadley cell on either side of the Equator, but several, which shift, merge, and separate over time in a complex manner.

POLAR CELL

Weather at high latitudes is dominated by a convection cell similar to the Hadley cell.

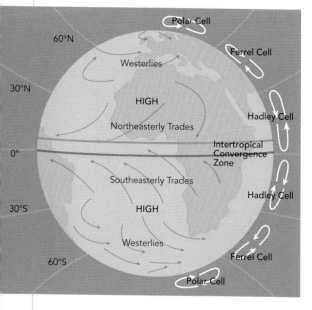

Above: Diagram shows Earth's three major convection cells in each hemisphere, the Hadley Cell, the Polar Cell, and the Ferrel Cell, also called the mid-latitude cell. Top left: The spiral quality of this low-pressure system over Iceland is a result of the Coriolis force, arising from Earth's daily rotation.

Although air at latitudes of sixty degrees north and south is cool and dry relative to tropical air, it is still warm and moist enough to rise into the tropopause and head poleward. By the time it reaches the poles, it is cold and dry and thus dense, so it descends to spread outward along the surface from the poles.

As a result of the Coriolis force, it also twists westward to produce the polar easterlies (winds from the east). The outflow of this dry, dense air from the poles also produces waves in the atmosphere called Rossby waves, which determine the path of the jet stream.

FERREL CELL

While the Hadley and Polar cells are truly closed loops, the Ferrel cell at the middle, or temperate, latitudes is not.

Although the net movement of surface air is from 30° to 60° in both hemispheres, the movement of upper level air is not well-defined. In part, this is because there is neither a strong heat source (as the thermal Equator is to the Hadley cell) nor a strong cold source (as the poles are to the Polar cell) to drive convection.

Moreover, the Ferrel cell falls at the latitudes that experience the strongest Coriolis force. Although winds are generally from the west (prevailing westerlies), they are at the mercy of passing weather systems, so they can vary sharply. For this and other reasons, the Ferrel cell is sometimes called the "zone of mixing."

A ball rolled straight across a spinning playground merry-go-round will curve at an angle relative to the merry-go-round's spin, an example of the Coriolis force.

THE CORIOLIS FORCE

A body in motion will tend to stay in motion in a straight line unless acted upon by an outside force. That's one of Newton laws. Roll a ball across a level floor and you'll see that this is true.

But roll a ball across a spinning playground merry-go-round, and its path will curve, causing the ball to fly off at an angle. This is because the merry-go-round is rotating as a solid body: All parts of its platform traverse the same angle each second, but that angle corresponds to a larger actual distance closer to its outer edge than it does near the center of rotation.

The same thing is true for Earth. Visualize looking straight down at Earth's North Pole. Every spot on the surface of Earth rotates once in 24 hours. Because Earth's equatorial circumference is somewhat more than 24,000 miles (38,600 km), a point on the Equator travels slightly faster than 1,000 miles per hour (1,600 km/h). But a point at the pole is rotating in place. Intermediate latitudes, such as New York City or Boston, are traveling at intermediate speeds—roughly 600 miles per hour (960 km/h).

Now, imagine firing a cannon from New York City due south toward the Equator. The cannon—and thus the cannonball flying out of it—has an eastward motion of 600 miles per hour. But points south are traveling eastward even faster, at speeds determined by their latitude (or, more precisely, their distance from Earth's rotational axis). So the cannonball—assuming it can actually fly as far as the Equator and get there in an hour—will land on the Equator not due south of New York, but some 400 miles (640 km) to the west. To hit a target due south of New York, the cannon actually would have to be aimed considerably east—a ballistics problem well-known and intensely studied by artillery gunners.

From the point of view of the person shooting the cannon, the cannonball would have appeared to curve to the right (clockwise in the northern hemisphere), as if it were acted on by a lateral force. Well, the force is real, and it is named the Coriolis force for a nineteenth-century French scientist. Moreover, the Coriolis force is a major factor in the direction of spin and path of travel of large-scale weather systems such as hurricanes (although not of small-scale phenomena, such as the direction water spins when going down a drain).

Global Winds

Earth's rotation and the resulting convection cells give rise to several important winds: the polar easterlies, the prevailing westerlies, the trade winds, and the jet streams. The first three are surface winds that have been important to mariners for centuries; the last are high-level winds first studied in detail in the 1920s, and now important to commercial airline pilots.

POLAR EASTERLIES

Although sometimes inconsistent, cool and strong polar easterlies form at high northern and southern latitudes as part of the surface return air flow of the Polar convection cell, radiating outward from the poles. When the polar easterlies meet warm, moist air that may be carried northward from the warm Gulf Stream current in the Atlantic Ocean, the combination can produce violent thunderstorms and even tornadoes in North America as far north as latitude 60 degrees.

JET STREAMS (WESTERLIES)

Jet streams are narrow bands of strong wind that form at the top of Earth's lower atmosphere, just below the tropopause. They flow generally from west to east at boundaries between warm and cold air. Year-round in both hemispheres, the main jet stream—the so-called "polar jet"—is found at the juncture of the Ferrel cell and the Polar cell at latitudes of around 50 to 60 degrees north or south, although the polar jet can meander from 30 to 70 degrees north or south. In winter, when there is a strong temperature difference between tropical air and polar air, a second

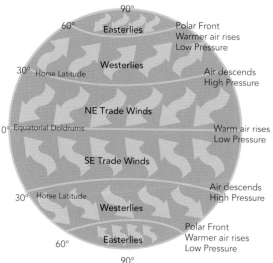

The global winds, shown with latitude indicators.

jet stream, the "subtropical jet," forms between 20 and 50 degrees north and south at the juncture of the Ferrel cell and the Hadley cell. Both jet streams vary in height from 4 to 8 miles, and can reach speeds exceeding 275 mph, although they also slow down and speed up along their rather meandering path.

Jet streams follow the movements of contrasts in air temperature, which roughly equates to following the Sun. Over North America, for example, the polar jet shifts north over Canada in the summer, but migrates south over the United States as autumn approaches, helping to bring

Above: As autumn approaches, the polar jet stream shifts south over the United States. Top left: A sailboat becalmed, a state often called "the doldrums." The term comes from the spot where the trade winds from both hemispheres converge, often become unsteady, and die out.

The Intertropical Convergence Zone, or ITCZ, where the trade winds converge, circles the Earth near the Equator, seen here as a horizontal band of cloud cover. Frequent thunderstorms take place in this zone.

down colder air. In fact, the path of jet streams steers cyclonic storm systems in the lower atmosphere; moreover, jet streams have a role in creating supercells, a type of storm system that give rise to tornadoes. Monitoring jet streams has thus become an important part of weather forecasting.

MID-LATITUDE PREVAILING WESTERLIES

The prevailing westerlies form at mid-latitudes (between thirty and sixty degrees north and south) as part of the surface return air path of the Ferrel cell. In both hemispheres the return is toward the poles, but because of the Coriolis force, the winds also have a net west-to-east direction. Because of all the complexities affecting the Ferrel cell, prevailing westerlies are less consistent winds than the trades, but they are responsible for moving much weather across the United States and Canada. Especially in the southern

hemisphere where there is less land to slow them, prevailing westerlies can be very strong, and are known as the Roaring Forties between 40 and 50 degrees south.

TRADE WINDS (EASTERLIES)

The trade winds, among the most reliable and consistent winds on the planet, are found in bands around Earth's equatorial regions. These are warm winds that early mariners used to blow their sailing ships from Europe to North and South America. Both north and south of the Equator, they are part of the surface return air flow for the Hadley convection cells from about 30 degrees north and south latitude toward the Equator. Because of the same Coriolis force that makes a cannonball fired due south actually land somewhat west, the trades also have a net east-to-west direction.

Where the trade winds from both hemispheres converge near the Equator, the air becomes

warmer and rises to feed the Hadley convection cell. Surface winds become unsteady and often die, forming a region of calm called the doldrums.

Palm trees bend in a tropical storm in Belize, which lies in the Intertropical Convergence Zone. Tropical storms can develop into hurricanes in the Gulf of Mexico and the Atlantic Ocean.

Tilt! Tilt! Earth's Seasons

Earth's axis of rotation is not perpendicular to the plane of its orbit around the Sun (the ecliptic). In fact, it is tilted at an angle of about 23½ degrees to the ecliptic. The astronomical direction in which the axis points (almost directly at Polaris, the North Star) remains essentially unchanged throughout the year. Thus, the amount of sunlight received at any location on Earth changes both in intensity and duration throughout the year, giving rise to the familiar four seasons of winter, spring, summer, and autumn, along with their attendant weather patterns.

SEASONAL PRIMER

Consider the first day of summer (summer solstice) in the northern hemisphere, June 21. That day, the north pole of Earth's axis is pointed most directly toward the Sun, and the Sun reaches the zenith at the latitude of 23½ degrees north, a fact that defines the latitude of the tropic of Cancer. North of the Equator, summer solstice is also the longest day of the year as measured from time of astronomical sunrise to astronomical sunset, or the sunrise and sunset times assuming Earth had no atmosphere. Conversely, in the southern hemisphere, June 21 is the first

day of winter, and the shortest day of the year.

At both the first days of spring and fall, which fall on or about March 21 and September 21, the Sun reaches the zenith over the Equator. Day and night are of equal length, so those days are called the equinoxes, from the Latin words meaning equal

Earth seen on the day of the vernal, or spring, equinox. On this day every year, the Sun's rays fall on both the Arctic and Antarctic regions at the same time.

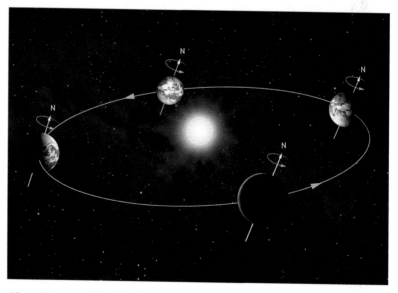

Above: Diagram of Earth's orbit, showing seasons arising from the planet's axial tilt. Top left: A surfer pictured in front of a setting Sun. Our view of the setting Sun is distorted by atmospheric refraction.

night. Vernal (spring) equinox in the northern hemisphere happens the same day as autumnal equinox in the southern hemisphere, and vice versa.

The first day of winter (winter solstice) in the northern hemisphere, on or about December 21, is the day the north pole of Earth's axis is pointed most directly away from the Sun, and the Sun reaches the zenith at the latitude of 23½ degrees south, the tropic of Capricorn. Of course, to those in the southern hemisphere, this day is summer solstice.

EARTH'S ORBITAL ELLIPTICITY

Earth's orbit around the Sun is not a perfect circle. It is actually an ellipse—a nearly circular ellipse, to be sure, but still the Sun is at one focus rather than being in the center of the orbit. So Earth is actually 3 million miles (5 million km) farther from the Sun when it reaches aphelion (farthest point in its orbit from the Sun) around July 4 than it is at perihelion around January 3.

So the sunlight falling on Earth's surface during northern summers is about 7 percent less intense than it is during southern summers. But that doesn't make northern summers cooler than southern ones. There is more land mass in the northern hemisphere than in the southern, and land absorbs heat more readily than oceans do. Thus, overall, northern summers actually tend to be somewhat warmer than southern summers.

In some ways, a sunset is an illusion. The Sun has already set, but atmospheric refraction raises its image from below the horizon to make it visible.

FUN WITH ATMOSPHERIC REFRACTION

Astronomically speaking, vernal and autumnal equinoxes—the first days of spring and fall on or about March 21 and September 21—are the two days of the year when planet-wide both day and night last an equal 12 hours.

But if you listen carefully to the times of local sunrise and sunset broadcast over local NOAA (National Oceanic and Atmospheric Administration) weather radio, you may hear that the days of equal day and night occur about a week later than autumnal equinox and about a week earlier than vernal equinox. In other words, year-round, every location on Earth seems to be getting a few minutes a day more sunlight than astronomers predict. How can that be?

There is a simple answer: Earth has an atmosphere. And that atmosphere refracts, or bends rays of light, just as water does. Stick a straw into a glass of drinking water: From the side, the straw will look bent. Sight down the straw, and you will see that the tip of the straw appears to be bent upward. The same thing happens with the atmosphere and the Sun. When the lower limb, or edge of the Sun seems to rest on the horizon, what you're seeing is just a refracted image of the Sun; the real Sun is actually completely below the horizon. Atmospheric refraction is pronounced enough to lift an image of the Sun a complete solar diameter. Thus, times of astronomical sunrise and sunset differ from terrestrial sunrise and sunset.

Atmospheric refraction also accounts for why the Sun looks squashed near the horizon: Lower, thicker layers of the atmosphere bend light more than higher, thinner layers, raising an image of the Sun's lower limb more than the image of its upper limb.

Possible Long-Term Astronomical Influences

Geological and paleontological evidence has revealed that Earth's climate has dramatically changed many times over the planet's history, with at least some changes apparently in cycles of tens to hundreds of thousands of years. Although understanding of long-term climate variations is incomplete, astronomers know of some long-term astronomical effects on our planet that could influence its future climate.

EARTH'S WOBBLING AXIS

Earth is not a perfect sphere. It also is elastic rather than solid all the way through, because its interior is molten. Thus, because it rotates rapidly on its axis, it

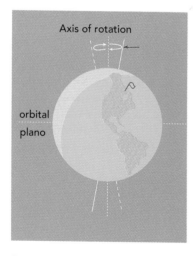

Diagram showing the wobble of Earth on its axis known as precession.

bulges out a bit at the Equator and is flattened a bit at the poles.

Tugging on that bulge is the Moon. The plane of the Moon's orbit around Earth is tilted about 5 degrees from Earth's equatorial plane; thus, for half its monthly orbit the Moon is north of Earth's equator, and the other half it is south. The Moon is so massive and close to Earth that its gravity alternates between attracting the bulge northward for two weeks, and southward for another two weeks.

Moreover, because Earth's rotational axis is tilted about 23½ degrees to the plane of its orbit around the Sun, to a lesser extent (because of the Sun's greater distance), the massive Sun also tugs on Earth's equatorial bulge.

The combined result of Earth's gravitational tug-of-war with the Moon and Sun is to make Earth act like a wobbling toy top: Earth's rotational axis describes a giant conical motion in space compared to its orbital plane around the Sun, one full circuit taking roughly 26,000 years. This large, slow wobble is called precession.

The effect of precession on Earth is to change the direction of true (astronomical) north—by a huge amount. Although this change is too gradual to be noticed by the average person,

Above: Artist's rendering of the Earth, Moon, and Sun. The Moon pulls on the bulge of the Equator. Top left: An infrared NASA picture of Saturn.

precession changes the positions of all the stars enough that astronomers must update star atlases about every 25 years. Even over the course of human history, it has changed which star has been the North Star.

Precession affects Earth's climate over time, although exactly how is a matter of some debate. Because precession alters the exact season Earth will be closest or farthest from the Sun in its annual orbit, precession likely increases or decreases the seasonal difference between summer and winter. Today, northern-hemisphere winters may be relatively moderate because they coincide with Earth's perihelion. But 13,000 years ago or 13,000 years from now, when precession has made perihelion occur during northern summer, the temperature differences between northern summers and northern winters may be more extreme.

CHANGING TILT OF EARTH'S AXIS

In addition to precession, Earth's axis does not maintain a consistent tilt of 23½ degrees to the plane of its orbit around the Sun. In fact, its tilt varies from as little as 21½ degrees to as great as about 24½ degrees, over a period of about 41,000 years. Like precession, this changing of the tilt of Earth's axis (technically called obliquity of the ecliptic) affects which regions on Earth will receive greater average solar energy than others. The climatological effects of all these changes has been preserved in deep sea sediments, glacial layers, tree rings, ancient corals, and other paleontological evidence.

CHANGING ELLIPTICITY OF EARTH'S ORBIT

As if the slow changes in Earth's axial tilt were not enough, Earth's orbit—now fairly circular—becomes more elongated and elliptical over periods of about 100,000 years, taking Earth farther from the Sun for greater periods of time. Those cycles, which correspond well to the timing of ice ages and interglacial periods, have been named Milankovitch cycles, after the Serbian mathematician Milutin Milankovitch who studied them in the 1930s.

A true-color image of Jupiter taken from the spacecraft Cassini. *Jupiter and Saturn, because of their enormous mass, can change the entire solar system's center of mass.*

PLANETS THROWING THE SUN AROUND

It sounds like science fiction, but it may be science fact: one astronomical factor that could affect Earth's climate might be the position of the Sun within the solar system.

Even though the Sun accounts for 99 percent of the mass of the solar system, the solar system's center of mass (barycenter) does not rest at the center of the Sun. That's because the massive planets Jupiter and Saturn are far away, and so possess 86 percent of the solar system's angular momentum—effectively, a long lever arm. As the gas giants move in their orbits, they change the position of the solar system's center of mass. Over a period of centuries, the solar system's center of mass migrates from somewhere inside the Sun itself to as much as 9 million miles away, a quarter of the way out to the planet Mercury. Effectively, that amounts to a slow-motion "throwing" of the Sun around in a complex trefoil pattern. Visualize the process as two children holding onto the ends of a rope that is wrapped around the waist of a parent; as the children run around the parent at different speeds and pull their ropes from various directions, they can make their parent involuntarily take steps to retain balance.

Such "solar inertial motion," as it's called, is speculated to mix gases within the Sun, or have other significant gravitational effects on the Sun's thermonuclear engine, with periods ranging from 179 to 2,400 years. Solar inertial motion is hypothesized to possibly account for the Maunder minimum and other lows in the sunspot cycle—and their resultant effect on Earth's climate.

It Came from Outer Space?

What killed the dinosaurs, behemoths who ruled Earth for 155 million years, from the Triassic through the Cretaceous Periods? Fossil evidence puzzled paleontologists for more than a century, suggesting that dinosaurs the world over came to an abrupt and mysterious end 65 million years ago. Worse, at the same time, three-quarters of the world's existing species of animals and plants, in the oceans as well as on land, also became extinct in what is sometimes called the "Great Dying."

Around 1980, physicist Luis Alvarez from the University of California at Berkeley along with his geologist son Walter Alvarez announced what was then a

Artist's rendering of an asteroid about to collide with Earth. It is believed that the asteroid that played a role in dinosaur extinction was between four and five miles wide.

dramatic new hypothesis: that Earth had been struck by an asteroid four or five miles across, which vaporized on impact

and threw vast quantities of dust into the atmosphere. For several years, the dust would have darkened Earth, reducing sunlight reaching the surface to only a fraction of its present intensity, killing off many of the green plants that needed bright sunlight to survive. Because many of the dinosaurs were vegetarians, they would have starved to death in a short time, leaving the carnivorous dinosaurs with no food source.

Scientists know that comets and asteroids have struck planets and other objects in the solar system thousands of times: witness the impact craters on the Moon. Even on Earth, several hundred ancient impact craters

Above: A diorama at the National Museum of Natural History, showing a scene from the Cretaceous period. The period ended about 65 million years ago, when scientists believe an asteroid slammed into Earth, triggering mass extinctions. Top left: Asteroid Ida, photographed in 1994 from NASA's Galileo spacecraft.

have been identified, the youngest and freshest (a mere 20,000 to 50,000 years or so old) being the mile-wide Barringer Meteor Crater in Arizona. Astronomers have even watched asteroid-sized objects striking other planets, the most recent being Comet Shoemaker-Levy 9, which broke into 21 fragments and plowed into Jupiter in July 1994. And climatologists know that ash and aerosols spewed from violent volcanic eruptions have affected Earth's weather worldwide for months or even years: after the eruption of the Indonesian volcano Tambora in 1815, the year 1816 was so frigid that it became widely known as "the year without a summer."

Today, geological dating and other evidence strongly suggest that Chicxulub Crater off Mexico's Yucatán Peninsula may be the very site of the asteroid impact that killed the dinosaurs. Chicxulub is about the right age: 65 million years old. It's also the right size; the impact of an asteroid the size of the island of Manhattan would blast out a crater some 60 miles across.

There is some ongoing scientific debate about that identification, as well as about the mechanism by which the dinosaurs died. For example, one hypothesis suggests that so much sulfur ejected high into the atmosphere would have caused sulfuric acid—essentially, battery acid—to rain back to Earth, poisoning plants and dinosaurs.

Still, it is pretty clear that Earth's historic climate has

The Chicxulub Crater on Mexico's Yucatán Penisula can be seen as a wide semi-circle of dark green on the upper left corner of the peninsula.

changed suddenly several times over the last quarter-billion years. Indeed, evidence of an ancient impact crater even larger (300 miles across) and older than Chicxulub under the Antarctic ice sheet in Wilkes Land suggests that another mass extinction between the Permian and Triassic Periods 250 million years ago—even before the Age of Dinosaurs—may also have been triggered by the impact of an even larger asteroid.

GUARDING AGAINST KILLER ASTEROIDS

If it happened once, it could happen again.

That's the philosophy motivating NASA Jet Propulsion Laboratory's Near Earth Object (NEO) surveillance program, whose automated telescopes are constantly photographing the sky, and astronomers analyzing the results, to catalogue and track every asteroid more than about 1 km (0.6 mile) in diameter. The purpose: to compute the orbits of any solar system objects that might be on collision course with Earth over the next century, and that would be big enough to cause cataclysmic physical destruction or climate change. Of course, the goal is to discover any such asteroid while it is far enough away from Earth either to deflect it from a threatening orbit, or to break it into smaller chunks that would be less destructive.

Lest one think that such a possibility is, well, astronomically rare, contemplate this: just in 2002 Earth had a close call, astronomically speaking, when a soccer-field–sized object sped by within 75,000 miles of Earth—only a third of the distance of the Moon. This NEO was the closest ever recorded for an object of its size.

CHAPTER 4

WEATHER VS. CLIMATE

Latitude plays an important role in climate. Left: Conditions within the Arctic circle, a region whose climate is characterized by year-round cold. Although they share latitude, the dry, hot Sahara Desert in northern Africa (top) and the verdant islands of Hawaii (bottom) have disparate climates. This is because climate is determined by an array of factors, including longitude, topography, wind, vegetation, and landmass.

Many people confuse the terms "weather" and "climate." Simply put, weather is what happens outside the window right now, but climate averages out what happens on this same date year after year, over the course of many years. Weather is what meteorologists report when they recount yesterday's rainfall; climate is the big picture, the long-term pattern of recurring weather.

Climate has a spatial dimension as well. Latitude plays a role—the climate of the Arctic differs from that of the tropics—but even regions at the same latitude can have widely differing climates; compare the parched Sahara with lush Hawaii. Moreover, the atmosphere strongly interacts with topography and vegetation on the land to produce distinct climate sub-regions, depending on local altitude, prevailing wind, bodies of water, humidity, and rainfall. Local climates are also heavily influenced by oceans.

Climate itself changes over time: Glaciers have advanced during ice ages and receded in warm periods. Climate has changed not just over geologic eras, but even within mere centuries. In addition to astronomical and other natural processes, humans increasingly are a major factor in climate change.

One Atmosphere, Many Climates

The world has dozens of climates, but many have basic characteristics in common. These common traits allow scientists to categorize climates by type. Although several systems are used, the classification scheme most widely seen in geography books is the one published in 1900 by the long-lived German meteorologist Wladimir Köppen (1846–1940).

CLIMATE ALPHABET SOUP

Observing that many climates at similar latitudes had similar patterns of temperature, Köppen categorized the world's climates into half a dozen major classifications. To each, he assigned a capital letter: **A** for tropical, **B** for arid (dry), **C** for warm temperate with a cool winter, **D** for cool temperate with a cold winter, **E** for polar, and **H** for highland.

Later, Köppen modified those major categories into sub-regions by adding a lower-case letter indicating a specific pattern for precipitation: **f** for wet year-round, **s** for having a dry summer season, **w** for having a dry winter season, and **m** for having a monsoon season (periodic major winds that often bring rainfall).

Köppen also added a third letter for indicating a specific

pattern of temperature: **a** for regions having a hot summer, **b** for a warm summer, **c** for a cool summer, and **d** for a very cold winter. In practice, only the dry (**B**), warm temperate (**C**), and cool temperate (**D**) climates get three letters.

Finally, he divided the polar climate (**E**) into two types: **EF** for regions that remain permanently frozen and covered with ice, and **ET**, for those that can support the growth of tundra.

Altogether, the Köppen classification system recognizes more than two dozen distinctly identifiable climates. Together, these climates tell a great deal about the physical characteristics of weather and climate at various latitudes.

LOW-LATITUDE CLIMATES

Corresponding roughly to the latitudes in both hemispheres dominated by the circulation of the Hadley cell, low-latitude climates are controlled by tropical air masses. Coinciding with the edge of the trade winds, they are centered on the tropics (Cancer and Capricorn) between 18 and 28 degrees north and south latitudes. They include tropical rain forests (Köppen classification **Af**), which are moist all year round, such as the Amazon River basin in South America and the

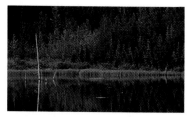

Above: Boreal forests are found in high-latitude climates; they are characterized by long cold winters and cool summers. Top left: The tropical savannas of the Serengeti in Tanzania have a wet season and dry season rather than a four-season year.

Congo River basin in Africa; savannah or grasslands (**Aw**) and tropical monsoon areas (**Am**), with wet summer and dry winter seasons; and true deserts (**Bw**).

MID-LATITUDE CLIMATES

Corresponding approximately to the unstable Ferrel cell, mid-latitude climates are affected by the conflicting polar and tropical air masses, neither of which can gain dominance. They include semi-arid steppe (**Bs**), both hot and cool, where moist ocean air is blocked by mountains, such as in the interior of the North American and Eurasian continents; the Mediterranean climate (**Cs**), with a wet winter and a parched summer, such as along the California coast and around the Mediterranean Sea; humid subtropical climates (**Cfa**) such as Florida, on the east coasts of

Although there are many climates on Earth, they can be broadly divided among six categories, some with subcategories based on precipitation patterns. Shown above, from top left (clockwise) are: arid, polar tundra, tropical rain forest, highland, cool temperate, and warm temperate (Mediterranean).

continents between 20 and 40 degrees in both latitudes; and the deciduous forest biome (**Cf**), the four-season climate typical of the eastern and midwestern U.S, southern Canada, northern China, and central and eastern Europe.

HIGH-LATITUDE CLIMATES

Dominated by the circulation of the Polar cell, high-latitude climates are found where Arctic air masses meet polar continental air masses at between 60 and 70 degrees north latitude over Canada and Siberia (no southern-hemisphere counterpart exists). Included are the boreal forest climate (**Dfc**), with long cold winters and cool summers, and the tundra climate (**ET**) along Arctic coastlines, with severe winters and a short mild season that is not really a summer.

HIGHLAND CLIMATES

At any latitude, the highland climate (**H**), cool to cold year-round, is found wherever mountains and mountain plateaus are above the tree line—the limiting altitude above which trees cannot grow, usually above 10,000 feet (3,000 m). Köppen assigned no subdivisions, even though local wet and dry seasons correspond to the pattern of the biome in which the alpine regions occur.

Land's Effect on Weather and Climate

Land affects local weather and climates in major ways, largely because of its thermal (heat-retention) properties, its altitude and topography, and its vegetation.

LAND'S WARMING AND COOLING

Under daily sunlight, land both warms faster and cools faster than water. Thus land next to an ocean or other large body of water is commonly subject to a diurnal weather pattern of onshore and offshore breezes familiar to sailors and bicyclists.

Typically, ground cools overnight, becoming chillier than any nearby large body of water; thus, at sunrise, air over the water warms, expands, and rises, drawing air from the land toward the water in a morning offshore breeze. Midday, when the temperatures equalize, the breeze may die down altogether for a few hours. In the afternoon, the Sun heats the land to be warmer than the water, and heated air rising from the land draws air from over the water toward the land: thus, the wind direction reverses, becoming an afternoon onshore breeze.

Above: On the island of Kauai, in Hawaii, Mount Waialeale (top) is the rainiest place on Earth, with an annual rainfall of up to 486 inches per year. Just 15 miles away, the dunes of Barking Sands (bottom) receive under 20 inches a year. This phenomenon is known as a "rain shadow," in this case created by Hawaii's towering volcanic peaks. Top left: The land area next to an ocean cools overnight, resulting in a morning offshore breeze.

LAND'S ALTITUDE AND TOPOGRAPHY

Mountains, hills, valleys, passes, and other topographic features profoundly affect local climates. For example, the volcanic peaks towering above 14,000 feet in Hawaii interrupt the steady flow of the humid trade winds. On the windward side, the mountains force the humid air to rise and cool, releasing its moisture in heavy orographic (mountain) rains. Moreover, the same mountains block rain from their leeward sides, literally casting "rain shadows." Thus, while the annual rainfall over the ocean around Hawaii averages just under 30 inches, Hawaii is home to the wettest spot on Earth— Mount Waialeale on Kauai, which receives up to 486 inches (1,234 cm) of rain per year; yet that sodden mountainside is only 15 miles (24 km) from Barking Sands dunes, which get fewer than 20 inches (50.8 cm) a year.

LAND'S VEGETATION

The transpiration (absorption and release) of moisture by plants significantly alters the temperature of local climates— something dramatically evident to any countryside bicyclist. Trees with their rich canopy of leaves retain the most moisture, so forests keep land coolest. Grasslands or prairies retain the least moisture and so are much warmer. Cultivated farmland falls in between.

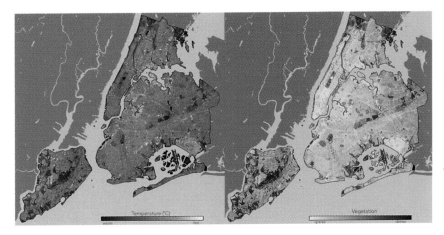

One example of an urban heat island is New York City. Compare the image on the left, which depicts the temperature range for the area with the image on the right, which shows vegetation for the same area. Notice the darker colored rectangular area near the center of each illustration: the cooler, tree-filled oasis that is New York's Central Park.

It is with vegetative cover that human hands show their effect on climate. Where humans chop down trees to plant crops, the local climate is warmed—but where humans have turned grasslands into cropland, the local climate is cooled.

URBAN HEAT ISLANDS

In cities, where humans virtually eliminate vegetation to erect buildings and pave streets, the local climate can become degrees hotter than surrounding land. Indeed, cities are literally urban heat islands, with bricks and pavement and concrete rapidly absorbing solar heat during the day and reradiating it at night—most evident during summer heat waves. Moreover, buildings are lighted and heated, releasing more heat into city air.

The urban heat island effect is of such concern that in New York City, the local electric company urges residents to plant shade trees and other vegetation on apartment roofs, balconies, and along the streets. The increased vegetation is intended to cool neighborhoods enough to lessen peak electrical demand for air conditioning.

Heat waves become even more unbearable in the underground microclimate of a New York City subway.

The Ocean Conveyor Belt

Ceaselessly circulating through all Earth's oceans is a worldwide loop of shallow warm ocean currents and deep cold ocean currents. Dubbed the ocean conveyor belt, it slowly and steadily distributes both solar warmth and ocean saltiness around the planet. In fact, because both temperature and salinity drive the conveyor, it is officially called thermohaline circulation (*thermo-* for heat and *haline* for salt).

First, surface ocean water warmed by the Sun flows from Equator to poles, carrying heat from the tropics to higher latitudes and releasing it into the air. The arm of the warm Gulf Stream that flows up the west coast of Scotland (the North Atlantic Drift), for example, is what keeps winters in the United Kingdom so mild, despite the fact that the British Isles are at the same latitude as frigid Newfoundland, Canada.

Second, ocean water cooled at high latitudes contracts, thus becoming denser, and sinks toward the ocean floor. Moreover, in regions so cold that icebergs form, such as in the far North Atlantic, the ocean water becomes so salty that it also gets denser and sinks (icebergs are made of fresh water, not frozen saltwater). Salinity also increases at the tropics despite frequent heavy rains, because tropical heat increases evaporation, leaving the salt behind in the water; this saltier, denser tropical water also sinks.

The net effect is a complex movement of currents circulating through all the world's oceans, perhaps taking up to 1,000 years for one complete circulation. Slow, but massive, the mass of water moving is enormous—some 100 times the volume of the Amazon River.

OCEANS AND CLIMATE

Tiny green marine plants called phytoplankton that live in the ocean are crucial to the web of life in the oceans, providing the food for many fish and even for giant mammals including whales. Through photosynthesis, the phytoplankton play an essential role—equal to all the vegetation on Earth's land masses—in absorbing the carbon dioxide that humans and other animals exhale, and converting it into breathable oxygen. Indeed, about half the oxygen in the atmosphere is produced by phytoplankton; by the same token, the phytoplankton help regulate

Above: Phytoplankton, seen here in a colored electron micrograph, are tiny marine plants that inhabit the topmost layers of oceanic water and regulate the amount of carbon dioxide in the atmosphere. Top left: As ocean currents, such as the Gulf Stream, flow from the Equator to the North and South Poles, water carries heat to the higher latitudes and releases it into the air.

the amount of carbon dioxide, an important greenhouse gas, in the atmosphere.

Cold ocean water also chemically dissolves carbon dioxide; the colder the water, the more it can dissolve, carrying it down deep into the oceans. Some of the dissolved carbon dioxide also becomes weak carbonic acid, of concern to biologists because greater acidity of ocean water can interfere with the growth of corals and the shells of marine creatures. Warm ocean water, by contrast, releases carbon dioxide into the atmosphere.

CLIMATIC CONCERNS

Scientists concerned about the possibility of global warming are closely monitoring the global conveyor belt. Measurements indicate that arctic glaciers,

especially those in Greenland, may be melting at a faster pace than in the past. As their fresh water runs off into the oceans, it would dilute the salinity of the salty water in the North Atlantic. The scientists wonder whether such dilution might be enough to reduce the density of the water, stopping it from sinking, and halting the global conveyor—including the Gulf Stream vital for warming Europe.

One of the counter-intuitive ironies of global warming is that it could freeze Europe. Indeed, the paleontological record indicates that the thermohaline circulation has halted in the past. The most recent dramatic event, known as the Younger Dryas, began suddenly 14,500 years ago. Greenland's mean

Icebergs, which are made of frozen fresh water, form in cold regions when ice breaks off of glaciers and falls into the ocean.

temperature fell a full 27° F (15°C), ushering in cold almost as glacial as during the last ice age. The Younger Dryas event also ended very suddenly 11,500 years ago, when temperatures over Greenland shot up again 15°F (10°C) in about a decade, possibly when the thermohaline cycle resumed. Today, some scientists calculate odds being as high as 70 percent that the Gulf Stream may stop flowing within two centuries, if no action is taken to reduce the amount of carbon dioxide humans are releasing into the atmosphere.

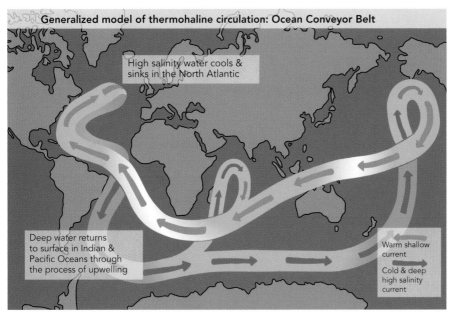

Generalized model of thermohaline circulation: Ocean Conveyor Belt

High salinity water cools & sinks in the North Atlantic

Deep water returns to surface in Indian & Pacific Oceans through the process of upwelling

Warm shallow current

Cold & deep high salinity current

Temperature and salinity are two forces that drive thermohaline circulation, also known as the ocean conveyor belt. This global circuit, which can take up to 1,000 years from start to finish, has a tremendous impact on Earth's climate.

The Ocean-Air Connection

In addition to the Hadley, Ferrel, and Polar cells that circulate warmer air north from Equator to poles and return colder air south across parallels of latitude, there are also several cells circulating air east and west across meridians of longitude. The most influential on climate is the Walker cell in the Pacific Ocean.

EL NIÑO–SOUTHERN OSCILLATION (ENSO)

As early as 1567, Peruvian fishermen noticed that something strange occasionally happened off the west coast of South America. Usually, nutrient-rich cold ocean water welled up from the ocean depths, and fishing was good. But every few years, usually around Christmas, the upwelling water was unusually warm; because the warmer water has fewer nutrients, less plankton grew, fewer fish fed, and fishing was poor. Because of its timing, by the nineteenth century the warm upwelling was dubbed El Niño (Spanish for the Christ child, when capitalized, or simply "boy child").

In the early twentieth century came an independent and seemingly unrelated discovery by British mathematician Sir George Thomas Walker (1868–1958), director of meteorological observatories in India. While most years India received seasonal prevailing monsoon winds with their life-giving heavy rains, every few years the monsoon was weak and India suffered drought and crop failures. Poring over the British Empire's long-time weather records collected from various parts of the world, Walker noticed that most years there was usually high atmospheric pressure recorded at sea level over the western Pacific Ocean (as measured from Tahiti) with low pressure recorded in the eastern Pacific (as measured from Darwin, Australia). But every few years, that pattern reversed: Higher pressure was recorded in the eastern Pacific and lower pressure in the west. Moreover, although the Pacific Ocean is halfway around the world from India, it seemed to Walker that the years of the reversed atmospheric pressure coincided with the years of weakened monsoons. Because the phenomenon seesawed back and forth across the Pacific, a later meteorologist called Walker's discovery the Southern Oscillation.

Today, it is known that the seesawing oceanic and atmospheric phenomena are physically related, so they are

Above: The warming effects of El Niño have caused cyclical weakening of India's monsoon rains. Every few years, rainfall is less than normal, causing drought and crop failure. Top left: Fishermen in the western Pacific such as these in Burma are affected by El Niño. It was Peruvian fishermen who first noted that periodically around Christmas, ocean waters became unusually warm, resulting in reduced yields.

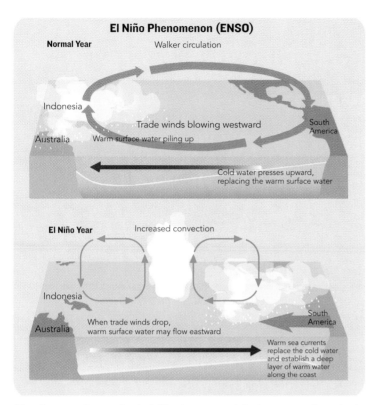

El Niño Phenomenon (ENSO)

Normal Year — Walker circulation

Indonesia
Australia

Trade winds blowing westward
Warm surface water piling up

South America

Cold water presses upward, replacing the warm surface water

El Niño Year — Increased convection

Indonesia
Australia

When trade winds drop, warm surface water may flow eastward

South America

Warm sea currents replace the cold water and establish a deep layer of warm water along the coast

Two diagrams show trade winds blowing westward during a normal year (top) compared to decreased trade winds during an El Niño year (above). Interactions between the ocean and the atmosphere in the tropical Pacific drive the El Niño–Southern Oscillation (ENSO) and cause an increase in surface temperatures as heat is transferred from the ocean to the atmosphere. When eastern trade winds decrease, warm sea currents extend eastward along the Equator.

now jointly called the El Niño–Southern Oscillation (ENSO).

THE WALKER CELL

The Pacific Ocean not only has the greatest mass of water of all the oceans, but also the broadest expanse across the Equator. Usually, the cold Humboldt current travels north along the west coast of South America, augmented by an upwelling of cold water off the Peruvian coast. In most years, the current flows west along the Equator and is warmed by the Sun, so the western Pacific near Indonesia is 3° to 8°C (12° to 14°F) warmer than the eastern Pacific off South America. In these normal years, warm, moist air over the western Pacific rises high into the atmosphere and flows eastward, sinking as cold, dry air over the colder eastern Pacific and driving the lower-level trade winds westward. That is normal Walker circulation: low-pressure air masses over Indonesia that bring rain, and high-pressure air masses over Peru that are dry, with strong easterly trades.

But every three to five years, for reasons still unknown, an El Niño warms the eastern Pacific off the coast of South America, until it is as warm as the western Pacific near Indonesia. El Niño disrupts Walker circulation, to the point where the southern-hemisphere easterly trade winds weaken and even reverse direction (become westerly), and the jet streams strengthen over the Pacific. Unusually high atmospheric pressure moves over Indonesia and Darwin, Australia, bringing dry spells and even drought, while low pressure moves over Tahiti and Peru, bringing heavy rains and even floods.

ENSO not only spans the Pacific Ocean, but also affects weather and climates worldwide. In North America, for example, El Niño winters tend to be unusually dry and mild in Canada and the northern United States, but unusually wet across the southern states. And, yes, Walker was right: during El Niño, the monsoon and rainfall lessen in northwestern India.

LA NIÑA

As often as El Niño warms the Pacific off South America, in other years the upwelling water is unusually frigid, strengthening normal Walker circulation and the trade winds—a phenomenon dubbed La Niña ("the girl child"). Today, El Niño is recognized as being the warm phase of the ENSO, and La Niña is recognized as the cold phase. In general, the global climatic effects from La Niña are opposite to those from El Niño, although complexities can alter the specifics.

Not Your Father's Atmosphere

The air we inhale today is not the atmosphere Earth was "born" with. In fact, it's actually the third atmosphere the planet has had.

Some 4.8 billion years ago, Earth's original atmosphere was composed of hydrogen, helium, and other cosmic elements in the original nebula from which the solar system was formed. But back then, the Sun was also much younger and hotter, blowing its residual mass into space in what is now called its T Tauri phase (named after a star in the constellation Taurus now observed to be doing just that).

This strong solar wind literally blew away much of Earth's original atmosphere.

After its formation, Earth was also hot. Many volcanoes spewed forth not only molten rock, but also such gases as carbon dioxide, water vapor, nitrogen, various sulfur compounds, and a lot of hydrogen as a reducing (hydrogen-based) atmosphere. Through complex chemical reactions, lightning, and other processes, much of the water vapor condensed to form oceans,

Cycads, an ancient family of evergreen seed plants that originated during the Jurassic period, are found in subtropical and tropical parts of the world.

while much of the carbon dioxide became fixed into carbonate rocks such as limestone.

Eventually, early life—primitive plants and bacteria—arose, which gave off oxygen as a waste product. For a billion years or so, the oxygen was largely taken up by iron on the surface and under the seas. Eventually, however, oxygen began accumulating in the atmosphere. It was then that Earth came to have an oxidizing (oxygen-based) atmosphere, along with a lot of nitrogen (which does not easily dissolve in oceans). This third atmosphere arose some 2.5 billion years ago, although the concentration of oxygen didn't reach today's 21 percent until sometime within the past 600 million years.

FIRE AND ICE

The temperatures on Earth today have not remained at current levels. For the first couple of billion

Above: Eroded limestone cliffs in Cabo Carvoeiro, Portugal. Earth's gaseous second atmosphere was hydrogen-based and set the stage for chemical reactions that ultimately created oceans and sedimentary carbonate rocks such as limestone. Top left: Earth's early atmospheres were greatly impacted by volcanic eruptions, which released gases such as carbon dioxide, water vapor, and nitrogen in addition to molten rock.

years after the planet formed, Earth was much hotter—heated by the frequent bombardment of large meteorites as well as by its own internal seismic and volcanic activity.

Scientists are able to study past climates by drilling and extracting deep cores from ocean beds or polar ice caps. These cores reveal climate periods in the past that were significantly warmer than the average temperature today. For example, there were hardly any polar ice caps during some of the time dinosaurs roamed the planet. Conversely, past climates have also been brutally colder, notably during the ice ages, when half of what is now the United States was covered with glaciers more than a mile thick.

How fast can the world's climate change? Studies show that there have been relatively sudden climatic shifts between the hot and cold extremes. A couple of climatic shifts and mass extinctions of plant and animal life have been traced to the impact of asteroid-sized objects (more than just the one possibly linked to the dinosaur extinction). And a few are attributed to the shutting down of the ocean's thermohaline circulation when an especially large glacier slid off the Canadian shield into the Atlantic Ocean. Such shifts in global climate, in fact, are what punctuate the ending of one geologic era and the beginning of another.

Today, humans live in the Holocene epoch, the name given to the time since the end of the last ice age some 10,000 years ago. All told, the Holocene has been comparatively balmy, except for the frigid Little Ice Age from about 1300 to 1850 (experts differ on the exact dates), whose coldest decades coincided with the Maunder minimum of a nearly complete absence of sunspots. Over the past 150 years, Earth's global temperature has risen steadily; in 2005, it reached the highest level in the last 1,200 years. Many scientists today are concerned that the rate of increase in our global temperature over the last century is unprecedented and cannot simply be the result of geological or astronomical processes—that it very likely is human-induced. If left unchecked, it could have grave consequences.

Brown sediment trapped in ice drilled from the frozen surface of Lake Bonney, Antarctica. Radar studies of such ice samples show that the lower surface of the ice melts and refreezes in thousand-year cycles, slowly bringing sediments from the surface to the water below.

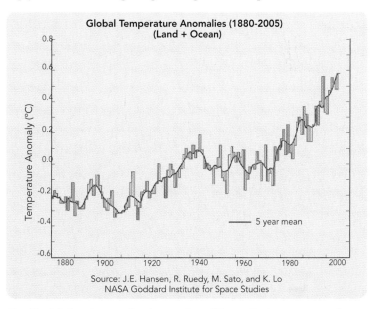

Earth's global temperature has increased steadily over the past century and a half. In 2005, temperatures averaged higher than they have in 1,200 years.

CHAPTER 5

GREAT ATMOSPHERIC CYCLES

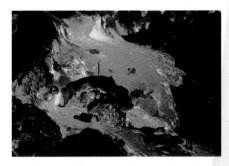

Left: Stromatolites in Shark Bay, Western Australia. The earliest known communities of life, stromatolites were the first to give oxygen off into the atmosphere. These cynano-bacteria form sticky mats of sand and sediment and created the possibility of life for all oxygen-dependent organisms. Top: Water molecules such as those present in lava at Volcano National Park in Hawaii are continuously recycled on Earth and journey through the atmosphere in many forms. Bottom: A refinery in Saudi Arabia. The environmental output of industry alters the chemical composition of the air, which (in turn) affects atmospheric cycles.

Inhale. Exhale. The air surrounds us, always there to breathe, yet we never run out of breathable oxygen. The entire planet is a self-recycling life-support system. The atmosphere interacts with the oceans, life forms, rocks, and the Earth's very interior to renew life-giving elements, compounds, and nutrients. Collectively, these never-ending interactions are called biogeochemical cycles.

Half a dozen planet-wide cycles are essential to life, to weather, and to climate. The big three are the oxygen cycle (which provides free oxygen in the atmosphere), the hydrologic cycle (which distributes water around the planet), and the carbon cycle (which moderates the Earth's temperature).

Atoms and molecules in these cycles are endlessly reused. A molecule of water might have burst from a volcano on a young planet Earth, then been breathed by a dinosaur an eon ago, to end up floating in the clouds of today.

Human activities such as agriculture, industry, and the destruction of wetlands and forests can alter the chemical composition of the air, and the availability and use of water, affecting all the atmospheric cycles. Even seemingly small-scale human actions can have far-reaching effects.

Open and Closed Cycles

Earth's atmospheric cycles are of two fundamentally different kinds: open and closed. Open cycles have inputs and losses outside the planet, whereas closed cycles completely recycle elements available on Earth.

EARTH'S OPEN ENERGY CYCLE

Earth is constantly bathed in radiation from the Sun. Thus, energy comes in from outside Earth. Some of it is absorbed by oceans and land or used by living beings; think of photosynthesis in green plants, which are then eaten by animals and humans. In a sense, when we eat an apple, we are eating Sun energy, converted for us by the apple tree.

Earth also radiates a fair amount of energy back into space. Sunlight bouncing off polar ice caps or snowfields, for example, is reradiated by the ice's reflective qualities. If Earth absorbed all the solar radiation it received, it would rapidly heat up like an oven.

The planet's energy cycle is open: energy Earth loses to space is also replenished from outside Earth—from the Sun. Indeed, without the Sun, all life and weather on Earth would cease; without the Sun, the atmosphere's own nitrogen and oxygen would freeze, blanketing the planet with solidified air.

EARTH'S CLOSED BIOGEOCHEMICAL CYCLES

While the Earth's solar energy is constantly replenished, the planet's chemical and nutrient

Above: A meteor shower, upper left. The brighter streaks of light are stars, which appear blurred through the use of time-lapse photography. Top left: Sunrise on Alpamayo mountain, Peru. Some of the solar energy that reaches the Earth bounces off snowfields and ice caps and is reflected back into space; this solar energy exchange is one of Earth's open energy cycles.

levels remain fixed. It is possible that a small amount of chemicals could arrive from outer space in the form of occasional meteorites, but that would be a rare source indeed. Earth's chemicals do undergo change, but only by being recycled through the planet's various processes and converting them into nutrients and waste products. In short, chemicals, including all the constituents of the atmosphere, operate on a closed system.

The atmosphere is cycled through systems in a number of ways: biological processes cycle it through living beings, geological processes cycle it through the lithosphere (rocks and Earth's interior), and chemical processes cycle the atmosphere through Earth's hydrosphere (oceans

and other bodies of water). Meteorologists and ecologists group all these complex and interacting cycles under the grand umbrella adjective "biogeochemical."

RESERVOIRS AND POOLS

All the biogeochemical processes operate continuously, but sometimes chemicals get stored for longer or shorter times in certain places. Scientists distinguish between two categories of storage. "Reservoirs" are long-duration storage—carbon stored as coal, for example. "Exchange pools" are short-duration storage; animals and plants, for example, store carbon for relatively short periods of time, in geological terms. Usually reservoirs are inanimate systems,

and exchange pools are living systems. A chemical's "residence" refers to the amount of time it is stored in one place.

Above: A historic photograph of a miner employed by the Consolidation Coal Company, which remains the largest coal producer in the United States. Carbon is stored for relatively long periods in coal deposits. Top: This achondrite from the Smithsonian's Gems and Minerals collection is a meteorite. Meteorites are rare examples of the Earth receiving chemical replenishment from outer space. Aside from meteorites, the Earth's biogeochemical cycle is closed.

The Water Cycle: Where Rain Comes From

The water cycle, one of the principle cycles that sustain life on Earth, is simple on a grand scale. Essentially, it consists of evaporation from the oceans and continents into the atmosphere, advection (horizontal transport) of water vapor by the atmosphere from oceans to land, precipitation over the entire globe, and runoff from land back into the oceans.

EARTH: THE WATER PLANET

All Earth's water in any of its phases or locations comprises the planet's hydrosphere. The hydrosphere includes the oceans, lakes, rivers, clouds, rain, snow, ice caps, underground water tables, fog, and even the imperceptible moisture (water vapor) in the air that surrounds us. Just to get a feel for how much water is in Earth's hydrosphere, contemplate some staggering figures.

Earth's surface area is 197 million square miles (510 million square kilometers). More than 70 percent of it—138 million mi² (362 million km²)—is covered with water. Two-thirds of the globe—125 million mi² (325 million km²)—is ocean. The volume figures are even more impressive. The total volume of water on Earth is some 360 million cubic miles (1.5 billion cubic kilometers), 97 percent of it in the salty oceans—by far the largest reservoir.

Only 3 percent of the hydrosphere is freshwater. More than three quarters of the freshwater is tied up as ice in polar icecaps and glaciers in Greenland, Antarctica, and elsewhere. The balance of freshwater is mostly groundwater stored in underground sediments and rocks. Visible freshwater in snowfields, lakes, rivers, streams, and rainfall accounts for less than 1 percent of all the freshwater on Earth; of that, the five Great Lakes in North America—collectively the largest body of freshwater on the planet—are the lion's share.

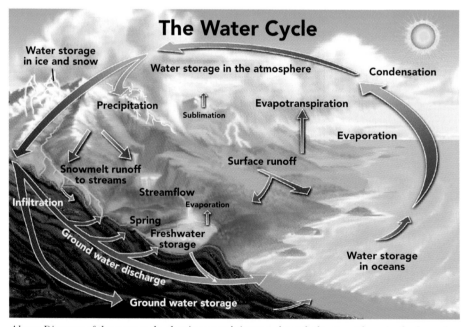

The Water Cycle

Water storage in ice and snow

Water storage in the atmosphere

Condensation

Precipitation

Sublimation

Evapotranspiration

Evaporation

Snowmelt runoff to streams

Surface runoff

Streamflow

Evaporation

Infiltration

Spring

Freshwater storage

Ground water discharge

Water storage in oceans

Ground water storage

Above: Diagram of the water cycle, showing water's journey through the atmosphere, in the form of ice, snow, rain, streams, rivers, lakes, surface runoff, groundwater, and oceans. Top left: Rain is an important part of the Earth's hydrosphere, distributing water over the Earth's surface.

Only some 0.001 percent of Earth's water is in the atmosphere, virtually all of it in the lower troposphere because of evaporation from oceans and land. Yet the atmosphere is essential for transporting water around the planet, especially from the oceans to the land.

Altogether, evaporation, transpiration, and precipitation cycle some 505,000 km^3 of water—nearly 22 times the volume of the Great Lakes—through the atmosphere each year. More evaporates from the ocean (434,000 km^3) than falls over the ocean (398,000 km^3). So a disproportionate amount—one and a half times the volume in the Great Lakes—falls over Earth's land masses, being transported there by the atmosphere.

POWER OF EVAPORATION

The Sun powers the water cycle. Specifically, solar radiation heats ocean water, causing it to evaporate—that is, change phase from liquid to vapor. Once in the atmosphere, water molecules remain airborne an average of 9 or 10 days, although they can return to earth after just a few hours or after several weeks. Evaporation from oceans as well as from seas, lakes, rivers, and other bodies of water provides 90 percent of atmospheric moisture. Evaporation from land accounts for the remaining 10 percent (see sidebar "Breathing Plants").

PRECIPITATION

Precipitation is condensed water vapor that falls to Earth's surface. Most falls as liquid rain, but some falls in solid form as snow, sleet, hail, or graupel. And some precipitation, such as fog drip, is simply condensation.

Once in the air, water vapor circulates within the atmosphere. As it rises, it cools. Airborne particulates such as dust act as condensation nuclei, allowing vapor molecules to condense and collect. These vapor molecules then form microscopic droplets of liquid water, or, at very cold temperatures, tiny ice crystals. At first, such condensation droplets or ice crystals are so small and lightweight that they cannot fall. Instead, they form clouds. But as the droplets and ice crystals circulate within clouds, they grow through collision and other processes to form larger droplets, snowflakes, or ice. Eventually, these droplets become heavy enough to fall as rain, snow, or other precipitation.

Icebergs from the Ilulissat glacier in Greenland. More than three quarters of the world's fresh water is stored in polar ice caps and glaciers.

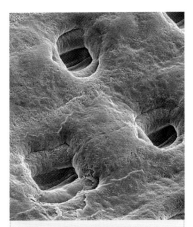

Electron micrograph of the surface of a Sitka spruce needle. The sunken pits are stomata, pores which are opened and closed by the two guard cells on either side of the depression (shown in light green). The spruce's stomata allow it to control the release of water vapor and other gases.

BREATHING PLANTS

Water evaporates from the ground as well as from oceans—witness a puddle drying after a rain. But over land, the single biggest source of water vapor is actually green plants. The trees and grass, the cornfields and potato patches are exhaling water even as they perform photosynthesis. In effect, plants give us water as well as food and oxygen.

Underground water is drawn up through the roots of plants. Other systems in the plant then spread out the water through the leaves, where it evaporates into the air through a process called transpiration. Anyone who has ever walked from a sunny clearing into a shady forest can actually feel the difference in humidity—and temperature—just from the presence of so many breathing trees.

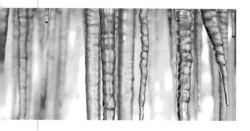

The Water Cycle: Transferring Heat

Water can exist in three different phases: liquid, solid (ice), and gaseous (vapor, or individual water molecules). Earth, unique among the planets of the solar system, contains significant amounts of water in all three phases. Depending on local temperature, water converts from one phase to another in the hydrologic, or water cycle. But that conversion from one phase to another also absorbs or releases heat.

Evaporation puts moisture into the atmosphere, but there's more to the picture. Water vapor also stores heat, allowing the atmosphere to transport heat from the Equator to the poles. Without this constant heat transport, we'd have no weather.

HEAT TRANSPORT IN THE MIST

When the Sun heats the oceans, water molecules at the surface absorb the heat, causing them to vibrate, or, in chemical terms, become excited. Some water molecules jiggle fast enough to break chemical bonds with other molecules and to escape as vapor that rises into the atmosphere. Thus, evaporation transfers heat from the liquid oceans to the escaping water molecules, which carry the heat away. "Latent heat" is the name given to such heat energy stored in molecules as a result of changing phase—in the above case, changing from water to vapor.

The energy transfer by latent heat has two important consequences. First, evaporation cools the oceans, just as evaporation of your perspiration during exercise cools your body. Such evaporative cooling is important to Earth's climate; without it, Earth would be far warmer than it is—an average of some 67°F (19°C) instead of its actual 59°F (15°C).

Second, evaporation transfers heat from the oceans into

	7.5
	6.0
	4.5
	3.0
	1.5
	0.0

Above: This diagram shows the variation in average atmospheric water vapor across the globe. Because of thermal differences, the atmosphere at the poles tends to hold less water than the atmosphere at the Equator. Top left: Icicles are formed by water in two different phases: liquid water drips from above and freezes as it travels downward, forming long, tapered columns of ice.

the atmosphere. Heat is then moved toward the poles through the circulation of the Hadley, Ferrel, and Polar cells. The latent heat is released only when the water vapor condenses (changes phase) back into liquid or ice. Thus, evaporation powers the circulation of the atmosphere.

Farther from the Equator, even more latent heat is released into the atmosphere again when water vapor freezes into ice crystals or other solid precipitation. Eventually, the snow or ice may melt, running into streams and rivers.

In short, water absorbs solar energy when changing phase from solid to liquid (melting) and from liquid form to vapor (evaporation). Conversely, water releases solar energy when it condenses from vapor to liquid to solid (freezing).

The physical process of absorbing, transporting, and releasing latent heat is tremendously important to climate. The amount of solar energy transferred from Equator to poles by latent heat is almost triple the amount that is carried by dry air.

Frozen carbon dioxide, or dry ice, here used to create smoke effects at a rock concert. Frozen carbon dioxide, like water, can transform directly from a solid to a gas in a process called sublimation.

SUBLIME HEAT TRANSFER

Although we tend to think of the normal progression of water through its phases as being from solid (ice) to liquid to gas (vapor), in fact water can also go directly from solid to vapor. That process is called sublimation.

Anyone who has ever received a shipment of meat or ice cream packed in so-called "dry ice" (frozen carbon dioxide) has seen solid carbon dioxide sublime into vapor. This is the explanation for the fog arising from the chunks of dry ice—in fact, the substance is called "dry" ice because it does not melt into a liquid first.

Water sublimes at temperatures below freezing, when ice can't first melt into a liquid. Indeed, sublimation of water is used commercially to freeze-dry foods, to purify chemicals, and to make refrigerator freezers frost-free. Because sublimation of water naturally occurs primarily in Earth's polar regions, it is not a significant source of atmospheric water vapor.

Water vapor also can condense directly into ice without first going through a liquid stage—a process called deposition. Indeed, that's what happens when frost forms overnight on grass. Whether water will condense into a solid or a liquid depends both on atmospheric pressure and temperature.

Earth is the only planet in the solar system where a significant amount of water exists in three different phases: liquid, gas (vapor), and ice. Changes in water from one phase to another always involve a release or absorption of heat.

The Oxygen Cycle

Without breathable oxygen, human life would cease. The presence of so much free oxygen (O_2), though, is very unusual, because oxygen is highly chemically reactive: it readily combines with atoms and molecules. This is another case in which the Earth is unique among all the planets in the solar system.

Although there are geological sources of oxygen, the single most important source of free oxygen present in the air is photosynthesizing green plants. In fact, without green plants, some scientists estimate that Earth's atmosphere would run out of oxygen in just 5,000 years. It's hard to form an exact picture of such a scenario, as the planet's cycles would be so profoundly different without any green plants.

PHOTOSYNTHESIS

Animals and people breathe in oxygen and exhale carbon dioxide as a waste product. Green plants, on the other hand, breathe in carbon dioxide and exhale free oxygen as a waste product. Specifically, plants take in water, carbon dioxide, and solar energy, and convert them to carbohydrates, or sugars—plants' energy source—and oxygen. Our best oxygen-producing friends are fast-growing young trees on land, and single-celled phytoplankton floating in the surface of the oceans.

Earth's plants are vital to the composition of the atmosphere. Moreover, they perform this oxygen exchange relatively quickly—in about 2,000 years. Each year, the world's plants consume about 1/2000th of the atmosphere's carbon dioxide, and produce about 1/2000th of the atmosphere's oxygen. Put another way, all the carbon dioxide and oxygen in the atmosphere today has been completely exchanged since the Common Era began. In contrast, it takes 2 million years for all the water in the oceans to circulate through the hydrosphere. Without plants, the water cycle would change somewhat, but the oxygen cycle would be drastically altered. Oxygen also gets cycled between the biosphere and lithosphere, although at a far slower rate. Living creatures in lakes and oceans make shells of calcium carbonate ($CaCO_3$), which is rich in molecular oxygen.

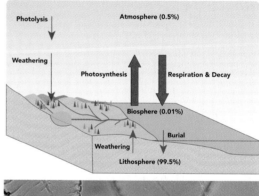

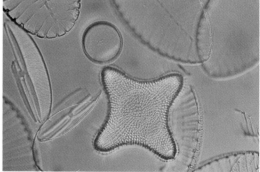

Top: This diagram shows the movement of oxygen through the atmosphere, biosphere, and lithosphere. Photosynthesis powers this cycle and makes oxygen available to animals and humans. Bottom: A micrograph of diatoms. Diatoms are single-celled phytoplankton, found in fresh and salt water, and are a major contributor to the atmosphere's available oxygen. Top left: Green photosynthesizing plants like this bamboo in Maui, Hawaii, take in solar energy, carbon dioxide, and water to produce oxygen.

When such creatures die, their shells fall onto the lake bed or sea floor, and are eventually buried, creating limestone rock. Though it seems astonishing, rocks have locked within them 100 times more oxygen than is present in the atmosphere. After untold millennia, plants absorb the nutrients from rocks, releasing the locked-up oxygen through photosynthesis.

THE OZONE-OXYGEN CYCLE

While plants on Earth are busy creating breathable oxygen for the entire atmosphere, another important chemical cycle is occurring in the stratosphere 20 to 40 miles (30–60 km) overhead: the ozone-oxygen cycle. When an oxygen molecule (O_2) absorbs energetic short-wavelength ultraviolet radiation from the Sun, it splits apart into two individual oxygen atoms ($2O$), which are very reactive indeed. Interestingly enough, when individual oxygen atoms (O) are in the presence of oxygen molecules, solar ultraviolet rays can make them combine into fragile molecules of ozone (O_3). Conversely, ozone itself is fragile; it is also readily split apart by solar ultraviolet into its constituent oxygen atoms and molecules. In other words, high in the stratosphere, oxygen atoms, oxygen molecules, and ozone constantly transform into one another—and all the reactions are powered by solar ultraviolet radiation.

Effectively, the ozone layer acts as Earth's sunscreen, blocking most of the Sun's energetic ultraviolet that would otherwise kill plant and animal life. The ozone layer is also important as it shields Earth's atmospheric water from being dissociated into its constituent hydrogen and oxygen, which would interfere with the water cycle, and all the life forms that depend on it.

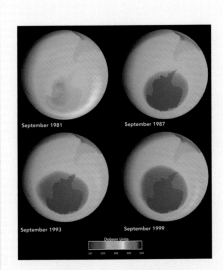

These images from NASA's Total Ozone Mapping Spectrometer (TOMS) show the deterioration of the ozone layer over Antarctica from 1981 to 1999. The ozone hole has been steadily increasing in size since the early 1980s due to ozone-destroying chemicals like chlorofluorocarbons and nitrous oxides released by combustion.

HOLES IN THE ATMOSPHERE

In the 1970s, British scientists discovered an appalling fact: each year the Earth's ozone layer develops a hole over Antarctica, where the amount of ozone is significantly depleted. To varying degrees, every year the hole has been growing both in area, reaching 11.4 million square miles (29.5 million km²) in 2006, and in duration, now lasting from August until early December. Some reduction in the amount of ozone is normal during the dark polar winters; today, however, over an area larger than North America, between 50 and 90 percent of the stratospheric ozone annually disappears. Scientists are now concerned that a similar hole may have started to develop over the Arctic.

In addition, over the past two decades, the general concentration of stratospheric ozone has been decreasing over temperate latitudes by about 3 percent per year. Ozone destruction is largely caused by synthetic chemicals used in modern devices. The greatest culprits are chemicals containing chlorine, notably a group called chlorofluorcarbons (CFCs) used in refrigeration systems, aerosol spray cans, and certain manufacturing processes. Other ozone-destroying chemicals include bromine and various nitrogen oxides released by combustion, such as aircraft engine exhaust.

The Montreal Protocol, an international agreement signed in 1987, greatly curtailed the production and consumption of CFC-containing products. Most industrialized nations signed the treaty, but the concentration of CFCs in the stratosphere may take decades to disperse. Thus, the ozone layer will continue to decline apace for many years before the destruction begins to reverse.

The Carbon Cycle and the Greenhouse Effect

In some ways, the carbon cycle is the mirror image of the oxygen cycle: humans and other animals breathe in oxygen and give off carbon dioxide, while green plants breathe in carbon dioxide and give off oxygen.

In other ways, though, the carbon cycle is a complement to the water cycle, because both carbon dioxide and water vapor are greenhouse gases, warming Earth and allowing life to exist.

CARBON AND ROCKS

Through photosynthesis, plants extract atmospheric carbon dioxide and use it to form carbohydrates and other organic compounds (which, by definition, are carbon-based compounds). When plants decay or are burned, most of their carbon dioxide returns back into the atmosphere. Bacteria and plants are deposited in marine sediments. Over geological eons, under great heat and pressure underground, some sediments are compressed into oil, natural gas, or coal. Thus, geological processes lock carbon dioxide underground in fossil fuels. Once the coal and oil are extracted and burned, however, the carbon dioxide reenters the atmosphere.

Underwater plants—those not exposed to air for photosynthesis—use carbon dioxide dissolved in water. Aquatic animals also use dissolved carbon dioxide and calcium to form shells of calcium carbonate ($CaCO_3$). After the death of such animals, their shells get compacted and cemented into limestone, which also stores carbon dioxide. In volcanoes, however, some of their carbon dioxide also returns into the atmosphere.

Altogether, Earth's rocks hold the equivalent of 70 atmospheres' worth of carbon dioxide. Venus, second planet from the Sun, is slightly smaller than Earth, but its surface atmospheric pressure is a bone-crushing 90 times greater precisely because most of its carbon dioxide is free in its atmosphere rather than locked up in its rocks.

THE LIFE-GIVING GREENHOUSE EFFECT

Carbon dioxide, although a minuscule fraction of the atmosphere by percentage, is a very powerful greenhouse gas. A greenhouse gas is one that is transparent to short-wavelength solar radiation (such as visible or near ultraviolet)

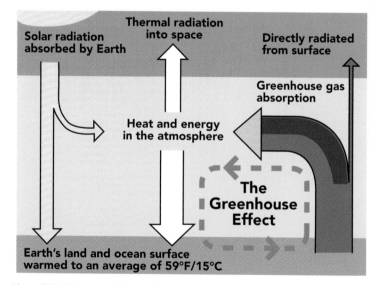

Above: This diagram of the greenhouse effect shows how solar radiation (energy) enters the atmosphere and is transformed into heat as it reaches the Earth. Greenhouse gases then trap some of the heat energy, preventing it from leaving the atmosphere. Top left: A fire blazes from the flare boom nozzle of an offshore oil rig. The burning of fossil fuels releases carbon dioxide into the atmosphere.

but absorbs longer-wavelength infrared (heat) that is given off by the warmed Earth itself. Thus, the presence of these gases slows the rate at which the planet radiates heat back into space.

Without the warm blanket of the greenhouse effect, most plant and animal life could not exist on Earth. The mean temperature of our planet would be a frigid 0°F (-7°C) instead of its actual balmy 59°F (15°C).

When the greenhouse effect is mentioned in discussions about global warming, the issue centers on human contribution to increased greenhouse gases. Since the beginning of the industrial revolution, unprecedented quantities of carbon dioxide and methane have been rapidly released into the atmosphere, through the burning of fossil fuels and other processes.

The shells of aquatic animals compressed over time created the limestone of these canyons in the American west. The formation of this kind of sedimentary rock can lock up carbon dioxide in the earth.

Greenhouse glass, like carbon dioxide, permits the transfer of solar radiation coming in, but prevents the release of heat out.

GREENHOUSE MISNOMER

Greenhouse gases came by their name through an early analogy to the way the transparent glass of a nursery greenhouse keeps tender lettuce seedlings warm even when February snow lies about. The comparison, however, is not wholly analogous.

A nursery greenhouse works by two mechanisms. First, glass transmits incoming visible wavelengths of sunlight, but is opaque to longer-wavelength infrared (heat) radiation. Thus, the glass admits more radiation than it allows to escape. In that sense, a nursery greenhouse is somewhat analogous to the atmosphere. The principal components of air (nitrogen and oxygen) transmit most sunlight down to Earth's surface, but trace amounts of greenhouse gases (carbon dioxide, water vapor, and methane) absorb most of the longer-wavelength reradiated upward by Earth's surface. Second, a nursery greenhouse blocks natural convection; the glass structure physically traps warm air, which readily escapes if a panel is opened in the roof. In this sense, the greenhouse analogy breaks down, because the so-called greenhouse gases do not block convection of Earth's atmosphere. Indeed, atmospheric convection actually carries both sensible warmth and latent heat from Earth's equator to its poles.

Are Humans Disrupting the Cycles?

During the last two centuries, humans have transformed the face of the planet. As populations have grown and nations have industrialized, they have deforested thousands of square miles and replaced them with paved highways, urban areas, and farmland. From the viewpoint of the atmosphere, these actions have removed sources of oxygen, and added sources of carbon dioxide. Moreover, industrialization has modified the heat balance and water cycle of the land; water evaporates faster from open farm fields than from forests, and water runs off asphalt rather than percolating through soil and underground water tables.

What effects might human actions have on global weather and climate?

HUMAN IMPACT ON THE WATER CYCLE

In the global water cycle, evaporation and precipitation remain poised in a delicate balance. If the world gets warmer, more moisture evaporates, causing greater precipitation. But this increase in the rate of the water cycle can be self-perpetuating: More airborne water vapor increases warming, thereby further accelerating the water cycle. Such a self-propelling

Above: Car traffic on a Massachusetts highway. The combustion of fossil fuels has contributed to a dramatic increase in the concentration of atmospheric carbon dioxide. This increase is thought to be a leading cause of global warming. Top left: The removal of forests increases carbon dioxide in the atmosphere through the destruction of photosynthesizing plants.

acceleration of the water cycle could result in more severe weather, although locally it could also create more cloud cover, causing local cooling.

Some scientists are apprehensive that recent greater severity and frequency of El Niños may stem from a water cycle that is being accelerated by global warming.

HUMANS AND THE CARBON CYCLE

The concentration of atmospheric carbon dioxide has steadily increased since the late 1950s from the burning of fossil fuels—coal, oil, and natural gas. But even converting wild acres to farmland increases carbon dioxide—and not just because forests and savannas are cleared. Frequent plowing of cleared areas breaks up soil, so decomposed organic material is not absorbed into humus particles, but is brought to the surface where its carbon dioxide escapes into the atmosphere or is washed away into rivers and finally into the sea.

Some fear that global warming from increasing releases of carbon dioxide might accelerate into a runaway greenhouse effect. As the Earth grows warmer, Siberian peat bogs are melting, releasing methane (CH_4), a carbon-based compound that is an even more powerful greenhouse gas than carbon dioxide.

The varied environmental effects of human activities—warming global temperatures, rising sea levels, isolated cooling in some areas—lead many experts in the field to refer to the issue as climate change rather than simply global warming. By any name, if the human-induced changes are left unchecked, the effects could spell disaster for the interconnected species that form the chain of life on this planet.

Above: A bulldozer moving logs in the South American rainforest. It is estimated that one-third to one-fifth of the world's carbon dioxide pollution comes from the destruction of tropical rain forests. Top: Siberian peat bogs are melting as a result of global warming. When temperatures rise enough to melt permafrost, the bogs release methane, a greenhouse gas more powerful than carbon dioxide.

CHAPTER 6

HOW WEATHER WORKS

Left: Hurricane Ivan, the ninth most power-ful Atlantic hurricane on record, caused major damage to Grenada, Cuba, and the United States and created a wall of water 131-feet (40-m) high, the tallest wave ever recorded. Top: The setting Sun shines on the people of the Masai tribe in Kenya, Africa. The rotation of the planet ensures that half of the Earth's surface is constantly absorbing solar energy while the other half radiates it back into space. Bottom: Cold winter storms in Canada are fueled by a complex interaction of meteorological systems including the movement of air flows, the water cycle, and local topography.

Earth's weather results from an interlinked chain of atmospheric cycles. Weather, and its constantly shifting nature, is as inevitable as the rotation of the Earth around the Sun, the spinning of the Earth on its axis, or day and night. In fact, these astronomical cycles drive changes in the weather.

On the day side of the planet, solar energy endlessly flows into the atmosphere, warming the oceans and land; the night side of the planet endlessly radiates this energy back. Because Earth is rotating, the energy-absorbing day side and the energy-radiating night side of the planet constantly change.

As the heat from the Sun warms the oceans and land, this warm air rises. Evaporation puts both moisture and latent heat into the atmosphere, which are carried by convection toward both poles. That, in turn, drives large masses of cold and dry polar air to flow back toward the Equator. Air flows are deflected toward the east or west by the Coriolis force linked to Earth's rotation, and gives rise to prevailing winds. Air flows and the water cycle are also affected by jet streams, topography such as mountains, and currents and upwelling of warm or cold ocean water. Also, human hands make a million small changes by replacing forests with farms and cities and changing the atmosphere.

Masses of Air

Ever wonder what the colored lines and arrows on TV weather maps indicate? Or exactly what meteorologists mean when they report that "a large mass of cold arctic air" or "a region of tropical air" will be moving into your area?

Weather is caused by the horizontal movement and conflict between huge masses of air having different temperatures and humidity. Knowing this, it is possible to read weather maps and get some feeling for the weather shaping up over the next few days.

FIVE TYPES OF AIR MASSES

An air mass is a large volume of air that takes on the characteristics of its source region—that is, the underlying environment over which it spends a substantial length of time, at least a day or two; hence the terminology "arctic" or "polar" air mass. To affect such a large body of air, a source region must be large, perhaps millions of square miles, such as an ocean or a continent.

Meteorologists classify air masses by humidity and temperature. Like the Köppen classification of climates, the characteristics of air masses are given letter designations that refer to their source regions. Approximate humidity is

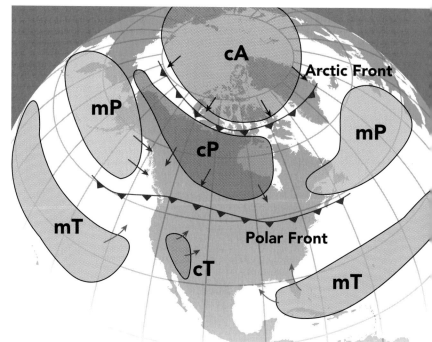

Above: Diagram of the source regions of air masses for North America: marine tropical (mT), marine polar (mP), continental polar (cP), and continental tropical (cT). Top left: Tropical air masses may originate in the Gulf of Mexico.

indicated by lower-case letters, and temperature range by upper case.

Humid air masses are classed as maritime or marine (m), because they usually originate over an ocean or other major body of water. Dry air masses are classed as continental (c) because they originate over land. Warm air masses are classed as tropical (T), cool ones as polar (P), and extremely cold ones as arctic (A). Although in theory there are six possible

combinations of letters, in nature there are only five types of air masses; maritime arctic (mA) masses do not exist because water freezes at arctic temperatures.

Weather in the United States is primarily affected by four of the five types of air masses: warm and humid marine tropical (mT) air moving north from the tropical Pacific Ocean, the Gulf of Mexico, and the tropical Atlantic; cool and humid maritime polar (mP) masses moving southeast from the northern Pacific or south from the northern Atlantic; and cool dry (cP) or arctic cold dry (cA) air moving south from ice fields and the land mass of Canada. Less influential is dry continental tropical air (cT) that can flow north from the Mexican plateau, although if it persists it can bring drought to the Midwest.

Air masses can dominate weather for periods ranging from under a week to an entire season. From June through August, for example, the Gulf Coast and the east coast of the United States commonly experience maritime tropical (mT) air masses, which causes their summers to be hot and muggy.

The United States, despite its large size, is itself not an

Clouds at the front of an approaching cold air mass bring high winds and a drop in temperature. Clouds often occur at the boundary between warm and cold fronts.

effective source region of air masses. Located under the relatively unstable Ferrel cell, the nation is frequently crossed by the passage of various weather systems; these systems disrupt any opportunity for air to stagnate

and take on the characteristics of the land below.

WHY WEATHER CHANGES

As air masses move around the planet, they take on additional characteristics of new areas underneath them. For example, in winter if a very cold and dry arctic mass (cA) were to move off Canada over the Atlantic, it would pick up warmth and moisture and become a maritime polar air mass (mP)—cool but moist. If instead the same cold and dry arctic mass were to move due south over the middle of the United States, it would gain warmth but not humidity, so would become a continental polar air mass (cP). The boundaries between air masses are known as fronts. Most dramatic or changeable weather takes place at a front.

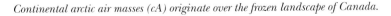

Continental arctic air masses (cA) originate over the frozen landscape of Canada.

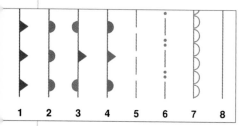

| 1 | 2 | 3 | 4 | 5 | 6 | 7 | 8 |

Fronts: Cold, Warm, and Otherwise

Fronts mark the boundary at Earth's surface where two large air masses meet. Fronts occur not just near Earth's surface, however; they also have a vertical structure. Because cold air is denser than warm air, it slides underneath warmer air, lifting it. That vertical movement leads to the formation of clouds and precipitation.

COLD FRONTS

The leading edge of a cold air mass displacing a warm air mass is called a cold front. The cold air, being denser, typically wedges underneath the less-dense warm air, pushing it vertically upward. Because cold fronts can move up to 30 mph (50 km/h), giving the cold air mass a steep slope, they can sweep into a region with surprising speed. The rising warm air cools, forming big, puffy cumuliform clouds (see chapter 7) that can grow into thunderheads, producing gusty winds, thunderstorms, and short-lived heavy precipitation that can include rain, snow, sleet, or hail. The air behind the cold front is usually drier and colder than the air it is replacing.

On a weather map, surface cold fronts are shown as a blue line with small blue triangles pointing in the direction the front is traveling.

Above: A long band of clouds formed by the collision of a warm and cold front. Top left: National Weather Service weather front indicators. Left to right: (1) cold front, (2) warm front, (3) stationary front, (4) occluded front, (5) surface trough, (6) squall line, (7) dry line, and (8) tropical wave.

WARM FRONTS

When a warm air mass displaces a cool or cold one, the leading edge is called a warm front. Again, because warm air is less dense than cold air, the warm front typically advances by sliding over the top of the colder air mass, but usually with a gentler, less steep slope. Warm fronts usually have more stable air (less vertical movement) so clouds tend to form in stratified layers. Farmers may welcome the steady, light, and widespread rain that warm fronts bring in spring and summer; in winter, though, warm-front precipitation can freeze into snow, freezing rain, or sleet. Temperatures may warm slightly, and winds are usually gentle.

On a weather map, surface warm fronts are shown as a line with red half-circles pointing in the direction the front is traveling.

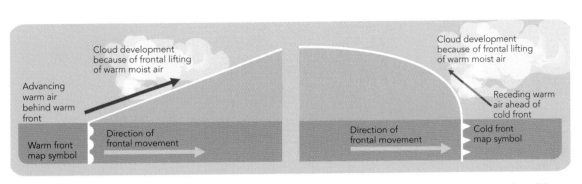

Above: Diagram of the movement of advancing warm and cold fronts. The image on the left shows how a warm front lifts and creates clouds as it advances towards a cold front. The image on the right shows a cold front pushing warm air upwards as it advances, also creating clouds.

STATIONARY FRONTS

If cold and warm air masses meet at Earth's surface but both are of roughly the same strength, the two may appear to stall over one location for hours or even days. Such stationary fronts are more common in summer than in winter. On a weather map, surface station-ary fronts are shown by a line with red half-circles alternating with blue triangles pointing in opposite directions.

OCCLUDED FRONTS

Occluded fronts arise from several causes. In all cases, however, three air masses are involved: cool polar air, colder polar air, and warm air. Where the two polar air masses meet, the colder one slides underneath the cooler one, blocking (occluding) the warm air from Earth's surface. The warm air is forced aloft above both polar air masses. Occluded fronts are important in the formation of mid-latitude cyclones (see chapter 9).

On a weather map, occluded fronts are represented by a line with purple half-circles alternating with purple triangles pointing toward the direction of travel.

Drylines, or the boundaries between more and less humid air masses, can result in violent thunder-storms like the one pictured here.

DRYLINES

Two air masses having only a minor temperature difference may still have a significant difference in humidity. A boundary separating warm humid air from warm dry air is called a dryline. Drylines, although rare east of the Mississippi River, are extremely important in the western and southern Great Plains of North America east of the Rocky Mountains, including regions in New Mexico, Texas, Oklahoma, Kansas, and Nebraska. Residents of these states are well acquainted with the violent thunderstorms and tornadoes that can form along drylines.

On a weather map, drylines are shown with a black line and open black half-circles pointing in the direction of movement

High- and Low-Pressure Systems

Because warm air is less dense—more buoyant—than cooler air, it naturally ascends. The rising air creates a slight partial vacuum, drawing in air from surrounding areas. When air is removed faster from the surface than it can be replaced by the inflow of surface air, the surrounding atmosphere becomes lighter in weight, thereby reducing its surface pressure (pressure, of course, being weight per unit area). A low pressure cell is like a mini version of convection cells (see chapter 3), air flows meet, the warm air rises, and cooler air rushes in to take its place. The

low pressure is created when the cool air can't keep up with the rising warm air. At the top level of a low-pressure cell, the warm air sent aloft then flows away from the upward-flowing column. The series of motions at work in a low pressure region are: together, up, and out.

Conversely, a high-pressure region is just the opposite. Cooler air is denser—less buoyant—than warmer air, so it naturally descends. If the air at the core of the high-pressure system is descending faster than it can flow away from the core at the surface, there is a temporary build up of the mass of air that literally

raises the weight (force) of the atmosphere in the column of the central core; the greater weight of the falling air is registered by barometers as increased air pressure. Thus, a high-pressure system has three characteristics just the opposite of a low: descending air currents at the system's core, diverging flows at the surface, and converging upper-level flows.

Because rising air cools, the moisture in it can condense. Thus, low-pressure regions are usually associated with clouds, precipitation, and unsettled weather. Because descending air warms, the moisture in it can evaporate. Thus, high-pressure

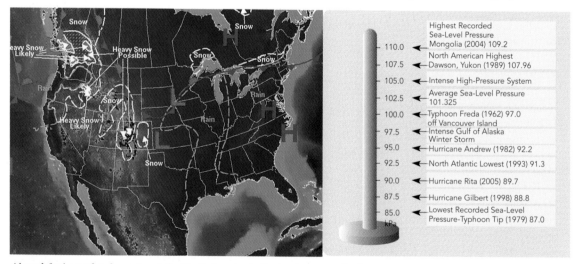

	Highest Recorded Sea-Level Pressure
110.0	Mongolia (2004) 109.2
	North American Highest
107.5	Dawson, Yukon (1989) 107.96
105.0	Intense High-Pressure System
	Average Sea-Level Pressure
102.5	101.325
100.0	Typhoon Freda (1962) 97.0 off Vancouver Island
97.5	Intense Gulf of Alaska Winter Storm
95.0	Hurricane Andrew (1982) 92.2
92.5	North Atlantic Lowest (1993) 91.3
90.0	Hurricane Rita (2005) 89.7
87.5	Hurricane Gilbert (1998) 88.8
85.0 kPa	Lowest Recorded Sea-Level Pressure-Typhoon Tip (1979) 87.0

Above left: A weather forecast map for North America indicates the center of a low pressure regions with a red L and the center of high pressure regions with a blue capital H. Above right: This chart shows the extreme range of recorded atmospheric pressures and its frequent association with intense meteorological events. Top left: Meteorologists let fly a rawinsonde balloon to measure atmospheric conditions including air pressure.

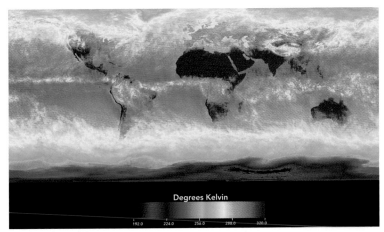

Above: This image from NASA's Atmospheric Infrared Sounder Experiment taken from the Aqua *spacecraft, shows the average temperature of the Earth's surface or any cloud between the space craft and Earth. The yellow band at the Equator corresponds with the ITCZ, an area of low pressure where thunderstorms and monsoons abound.*

regions are usually associated with clear, dry, stable weather.

On weather maps, the center of a low-pressure system is generally indicated by a red capital L, and the center of a high by a blue capital H.

SEMI-PERMANENT HIGHS AND LOWS

Although high-pressure and low-pressure systems commonly form, migrate hundreds or even thousands of miles, and dissipate according to sunshine, prevailing winds, topography, and other factors, there are regions on Earth where highs and lows persist for months or longer.

One of the most persistent low-pressure regions, for example, is the equatorial intertropical convergence zone (ITCZ). There, evaporation powers the Hadley cells both north and south of the Equator, humid air is constantly rising, cooling, condensing, and forming clouds favorable to the heavy rain showers of tropical afternoons. The ITCZ is a persistent band of low pressure, or more specifically, a band of individual convective storms—each a low-pressure center—that girdles the planet parallel to the Equator.

Where air in the Hadley cells sinks toward the surface around 20 or 30 degrees north and south latitudes, large bands of high air pressure also girdle Earth in what are known as the subtropical highs. As the descending air warms, its relative humidity drops, suppressing cloud formation and precipitation. Thus, it's no accident that the world's greatest deserts, including the Sahara and the desert southwest of the United States, are found in the subtropics, especially inland away from moisture from nearby bodies of water.

At the poles, where frigid air is descending, regions of high pressure persist that are known as the polar highs.

Other latitudes are dotted with a series of alternating semi-permanent cells of high and low pressure, whose positions and intensity alter with the seasons. In the northern hemisphere winter, the best-developed cells are the Aleutian low over the Pacific, the Icelandic low over the Atlantic, and the Siberian high over central Asia. In northern summer, the most prominent features are the Hawaiian high of the Pacific, the Bermuda-Azores high of the Atlantic, and the Tibetan low of southern Asia.

The Sahara Desert is located in a subtropical high zone. In this region, descending air warms, decreasing relative humidity, which suppresses both cloud formation and precipitation.

What Causes the Wind?

Wind is simply the air in motion. Usually when people mention wind, they are referring to horizontal motion. For example, a forecast of west winds of 10 to 20 miles per hour (16–32 km/h) means that horizontal breezes are coming from the west at speeds of between 10 to 20 miles per hour (16–32 km/h).

Wind direction and speed result from three forces: pressure gradient force, Coriolis force, and friction with the ground.

PRESSURE GRADIENT FORCE

Think about blowing up a balloon. After deeply inhaling air into your lungs, you exhale into the balloon. The balloon inflates because you are blowing into it with greater force (air pressure) than the surrounding atmosphere exerts on the balloon. When the balloon is nearly fully inflated, however, it becomes harder to inflate because

its internal air pressure becomes higher than the pressure from your lungs. If you stop inflating it and leave the end open, the balloon will completely deflate by allowing air to rush out, equalizing its internal air pressure with that of the rest of the room.

The whole process (minus any squeezing from the elastic rubber or from your lungs) illustrates how air will flow naturally from regions of higher pressure to regions of lower pressure. That also happens with the larger atmosphere: air naturally flows from high-pressure systems to low-pressure systems. That flow of air is wind. Some winds can be very strong. In that case, the difference in air pressure between neighboring highs and lows is great. On a weather map, surrounding H's and L's that mark the high- and low-pressure centers,

Above: The increasing difficulty of blowing air into a nearly fully-inflated balloon results from the pressure difference between the air inside the balloon and the air inside one's lungs.

Above: On a weather map the lines surrounding high- and low-pressure areas (marked red H and blue L) are called isobars. They delineate increments of increased or decreased pressure, similar to contour lines on a topographical map. Where the lines are close together, the pressure gradient difference is high, indicating strong winds. Top left: Humans have harnessed the power of wind for everything from wind turbines to wind surfing.

are lines of equal pressure called isobars—"iso" means "equal," and "bar" is a unit of pressure (see chapter 2). Isobars are analogous to the contour lines on topographic maps, which indicate lines of equal altitude above sea level. Where contour lines are far apart, a slope is gentle; but where they are packed close together, a hill is quite steep. The same is true of isobars: where isobars are packed together, the gradient in air pressure is great, and winds will be strong.

The difference in force between high-pressure and

low-pressure areas is called the pressure gradient. The pressure gradient force (Pgf) is a force that tries to equalize those pressure differences, by pushing air from high pressure toward low pressure. Indeed, in the absence of any other force, wind would flow in a straight line from high to low pressure.

CORIOLIS FORCE AND FRICTION

As seen in chapter 3, because Earth rotates as a solid body, the Equator is spinning faster (about 1,000 mph) than higher latitudes. Thus any long-range projectile is deflected to the right in the northern hemisphere (to the left in the southern hemisphere), a problem well known to artillery gunners. Because high- and low-pressure systems are so huge—commonly hundreds of miles across—the Coriolis force diverts the direction of wind flow. The Coriolis force is, indeed, what gives rise to prevailing winds, such as the trade winds. It is also what causes weather systems to travel in prevailing directions, such as across the United States from west to east. Within and between individual weather systems, another factor is also important: friction between air and Earth's surface. The topography of land and the presence of trees, buildings, and other irregu-

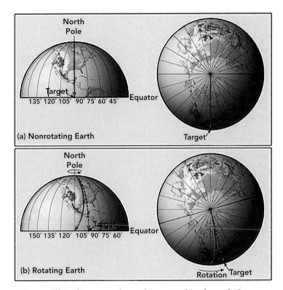

Above: This diagram shows how an object's path from the North Pole to the Equator appears to curve due to the Earth's rotation, a phenomena known as the Coriolis effect. The upper section of the diagram shows the object's trajectory if the Earth was still. The bottom section shows the object's intended path towards a target with a dotted line, and the actual trajectory with a solid red line.

larities slow the wind closer to the ground; think of treetops swaying in the wind when only a light breeze is felt on the ground. Thus, wind at ground level is slower than air flowing hundreds or thousands of yards above the surface. In addition, the pressure gradient force between highs and lows is greater higher in the troposphere, so upper-level winds tend to travel faster than lower-level winds. The combination of the Coriolis force, surface friction of Earth, and altitude differences in pressure gradient force tends to make high- and low-pressure weather systems spin in gigantic cyclonic patterns readily seen in satellite photographs of Earth.

Above: Wind turbines on the Oregon-Washington border. A day's wind, at an average of 45 miles per hour (72.42 km/h), can produce enough power for 70,000 homes.

Cyclonic Systems

Satellite photographs show late summer and early autumn hurricanes to be magnificent spinning whorls that are hundreds of miles across. But hurricanes are just one type of whirling weather system known as a cyclone—and winter storms are cyclones, too.

CYCLONES

Closed low-pressure areas are called cyclones. At Earth's surface, rising warm air at the core is drawing air into the cyclone from regions outside the low-pressure area. In the northern hemisphere, a combination of the Coriolis force and friction with Earth's surface makes cyclones spin counterclockwise (clockwise in the southern hemisphere). Because the rising air cools and forms clouds, classic examples of cyclones are intense tropical storms that bring strong winds and heavy rains.

Cyclones go by different names, depending on where they occur. In warmer months, cyclones are known as hurricanes in the Gulf of Mexico, the Atlantic, and the western Pacific; typhoons in the eastern Pacific; and simply as cyclones in the Indian Ocean. In all cases, they are powered by the latent heat released when water vapor condenses to form clouds. Once such storms move off warmer surface waters and migrate over land or colder waters, their sustaining heat source is cut off and they begin to lose strength.

WINTER CYCLONES

Cyclones are not just warm-season events. Winter storms that cross the United States from west to east are also cyclones. Such winter cyclones repeatedly follow distinct tracks across the United States and Canada

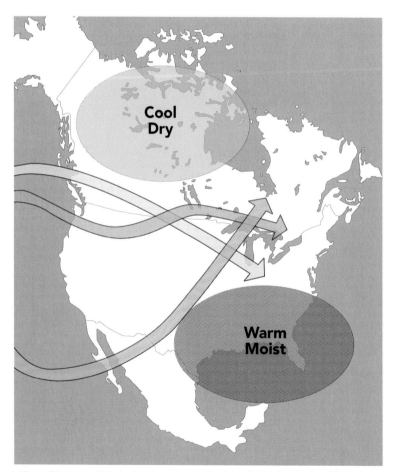

Above: Diagram of the paths that storms take through North America. The Alberta Clipper Storm is shown in blue, the Colorado Low in green, and the zonal pattern in pink. Top left: A thunderstorm swirls in a cyclonic updraft above a highway.

because of the configuration of the mountains and bodies of water. These cyclones are named after the place they originate.

One recurring winter storm track is the Alberta Clipper, which enters North America from western Canada but strengthens over the province of Alberta. Then, following the polar jet, it races south into the United States. The Alberta Clipper brings down frigid temperatures, but is so dry that it usually brings only light to moderate snow. Occasionally, sheer momentum can carry an Alberta Clipper off the mid-Atlantic or New England coast; when it meets relatively warm, moist air, its low-pressure center can dramatically deepen and dump heavy snowfall on the east coast.

Another favored area for spawning new winter storms is the eastern edge of the Rocky Mountains. Usually these storms are somewhat warmer and contain greater amounts of moisture. Storms developing over Oklahoma (Oklahoma Hookers) or Colorado (Colorado Lows) track north across the Ohio Valley or southern Great Lakes toward central or northern New England, dumping heavy snows on Chicago and the Midwest. Storms that develop farther south in Texas, notably the Texas Panhandle Low, often track east and turn north up the Appalachians, bringing heavy snows to Louisville and Cleveland. If they meet cold polar high pressure over the northeast, a secondary low can form off the East Coast. Winds shift to the northeast, keeping the cold air in place and dumping heavy snows on the Eastern seaboard as a classic Nor'easter.

Residents of Manila seek shelter from typhoon Xangsane, which struck the Phillipines and Vietnam in September, 2006. Winds raged at up to 140 miles per hour (225 km/h), caused hundreds of deaths, and at least 747 million dollars in damage.

After wreaking havoc on the village of Manchester, South Dakota, this F4 tornado plows through the neighboring farmland. Tornadoes are formed within cyclonic systems, but the terms cyclone and tornado are not interchangeable.

WHAT'S NOT IN A NAME

The term "cyclone" can be confusing. Hurricanes and violent tropical storms are sometimes called cyclones, and in the early twentieth century, the word "cyclone" was often used as a synonym for "tornado." People used to flee to so-called cyclone cellars or cyclone caves next to their houses as shelter from twisters.

Nonetheless, in meteorology today, the word cyclone refers to any closed low-pressure system that rotates counterclockwise in the northern hemisphere. Indeed, even a mild low that produces nothing more than a few clouds and gentle breezes is a cyclonic system.

Five Types of Global Cyclones

Worldwide, meteorologists generally distinguish between five main types of lows: tropical lows, extratropical lows, subtropical lows, polar lows, and thermal lows. In the first four, the adjective describes the geographic region where the low has formed. But they also differ in their structure and method of formations. In contrast, thermal lows can exist in the subtropics and at mid-latitudes.

TROPICAL LOWS

Tropical lows originate near the ITCZ about a dozen degrees north and south of the Equator, forming over the warm oceans and land masses. Because surface air is converging aloft, intense thunderstorms can begin organizing themselves into a rotating system that can slowly spin out of the tropics. Although such tropical lows may disperse, they also can strengthen into full-fledged tropical storms, eventually growing into hurricanes or typhoons. Tropical lows, which typically do not have fronts, are known as warm-core systems. These systems are formed by warm air rising in the absence of any cold air.

EXTRATROPICAL LOWS

Sometimes referred to as mid-latitude lows because they occur between 37 and 60 degrees north and south, extratropical lows form over much of the United States, Europe, and central Asia, and in the southern hemisphere, over Argentina, Chile, and portions of Australia. These cyclone systems form near the polar front, which drives the polar jet, separating cold, dry polar air from warm, moist subtropical air. Extratropical lows are known as cold-core cyclonic systems. They can last for days and can affect entire continents. Extratropical lows can result in precipitation ranging from scattered showers to torrential rains.

SUBTROPICAL LOWS

Subtropical lows form where tropical lows and extratropical lows collide. They can occur widely in latitude; classic subtropical cyclones are the Kona lows that batter Hawaii, but at least one subtropical cyclone

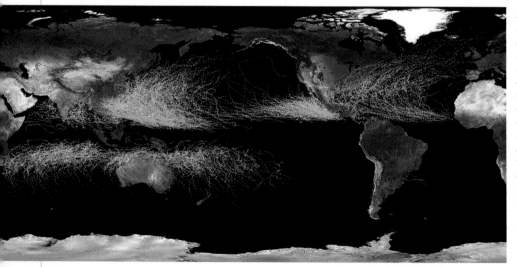

Left: The paths of all the tropical storms in the period from 1985 to 2005 measured at six hourly intervals. The strength of the storms is indicated by colors that correspond to the Saffir-Simpson scale. Top left: Zabriskie Point in Death Valley is a source region for thermal lows, systems that develop where the ground or water temperature is warmer than the air above.

has even formed over the Great Lakes (so-called Hurricane Huron on September 1996). Their common characteristic is that they form over warmer waters beneath a cold, upper-level low that has been cut off from prevailing westerly winds.

POLAR LOWS

Short-lived but intense, polar lows are capable of generating gale-force winds. These relatively small cyclones develop over high-latitude, mostly ice-free waters such as the northern Labrador Sea and the Gulf of Alaska, in addition to the waters surrounding Antarctica. The temperature difference between frigid Arctic air and comparatively warm waters creates a system that is sometimes referred to as an "arctic hurricane." The cyclones range in lateral size from 60 to 300 miles (100 to 500 km).

THERMAL LOWS

Thermal lows are stationary low-pressure systems that form over regions where ground or water is much warmer than the overlying air, such as over Death Valley in summer. Hot air rises, drawing in surrounding cooler air, and the system begins to rotate. Shallow in vertical depth compared to other lows, they decrease in strength at night and increase again during the day under the beating Sun. If a thermal low is very large, it can affect weather in surrounding regions. For example, a huge thermal low that forms over the Gobi Desert sets up prevailing winds that usher in the rainy Asian monsoon season.

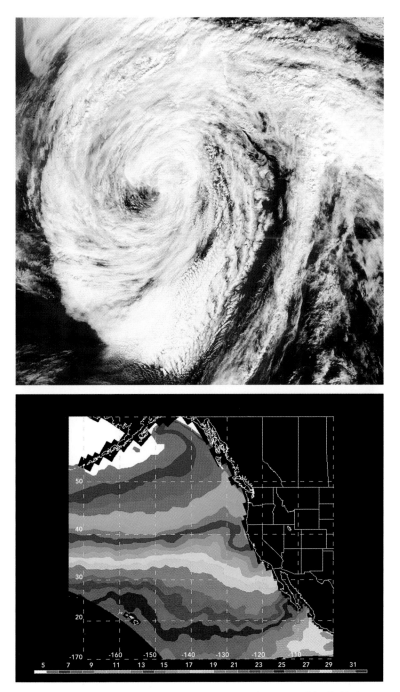

Above: This chart shows the range of temperature in the waters of the Gulf of Alaska. Although quite far north in latitude, these waters are relatively warm. The juxtaposition of warm waters below and cold air above creates polar lows.
Top: Hurricane Florence, originally a tropical storm, moved as far north as the Gulf of Alaska. In the process, the storm changed from a warm to a cold core system and became an extratropical storm.

Anti-Cyclonic Systems

Closed high-pressure areas are called anticyclones. In an anticyclone, air is flowing outward from the high-pressure center. Thus, the Coriolis force makes anticyclones spin in the opposite direction to cyclones: clockwise in the northern hemisphere, and counter-clockwise in the southern. Anticyclones are very stable, so they can persist for extended periods of time.

Although high-pressure areas are generally associated with clear, calm weather, anti-cyclonic systems are not always benign. In the United States during winter, southward or southeastward-moving cold fronts preceding an anticyclonic system can bring a sharp cold snap of continental polar (cP) air with arctic temperatures. They are also often a key culprit in droughts. Anticyclones also can usher in several types of hot, descending winds that have been given regional names.

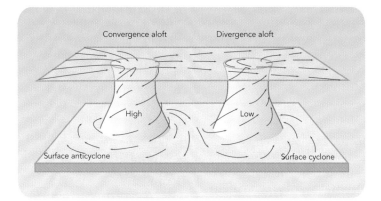

Above: Diagram of the different air movement patterns that form cyclones and anticyclones. Top left: Firefighters poised to battle the wildfire driven by the anticyclonic Santa Ana winds. In these systems, the air masses descend and compress, which causes strong winds and temperatures topping 104°F (40°C).

Air cooled over ice sheets on high plateaus in Greenland becomes extremely cold and dense. Gravity pulls this air downward creating katabatic winds.

SANTA ANA WINDS

Santa Ana winds are hot, dry winds over southern California common in the fall, and to a lesser extent in the spring, that originate from anticyclones over the Rocky Mountains or Great Basin. In the Rockies, air flows down the mountains' western slopes, warming and drying as it travels. Common wisdom holds that the Santa Ana winds are heated as they pass over deserts, but this is not actually the case. They indeed gain in heat as they travel, sometimes increasing in temperature by up to 54° (30°C), but this is caused purely by compression, as the winds meet higher atmospheric pressure at lower altitudes. Near the coast, Santa Ana winds cause temperatures to top 104°F (40°C), hotter even than such interior desert locations such as Las Vegas, Nevada. Santa Ana winds also have contributed to the rapid spread of destructive wildfires in Southern California, some of which have engulfed hundreds of square miles.

THERMAL HIGHS

A thermal high is the high-pressure mirror image of a thermal low—that is, a shallow high-pressure system that results from the chilling of the surface below. Like a thermal low, a thermal high remains fixed over one location.

In its purest form, air chilled over a high plateau—such as the Antarctic ice sheet and the Greenland ice sheet—becomes exceptionally cold and dense. As the cold air increases in density, gravity causes it to flow down the slopes of the plateau. Such downward flowing winds are called katabatic winds, from the Greek words for "going downhill." They continue flowing until the upper-level cold air is depleted. Although in most places katabatic winds are only light breezes, when funneled through narrow, steep canyons they can reach speeds of between 60 and 120 miles per hour (100 to 200 km/h). For that reason, the spot on Earth subject to the highest average wind speed year-round is Cape Denison in Antarctica.

Other spots on Earth also have katabatic winds, often going by local names. For example, those in France flowing out of the Alps into the Rhone River Valley are called mistrals, and those flowing out of the Balkan Mountains toward the Adriatic coasts are called boras.

A coral pink chinook arch spreads across the sky. Chinooks can change local temperature from 32°F to 100°F (about 0°C to 38°C) in the course of a morning.

CHINOOKS AND FOEHNS

Although driven more by low-pressure than by high-pressure systems, unlike the Santa Ana and katabatic winds, there are other winds that flow down mountain slopes and warm by compression. Most prevalent in winter, and ushering in unseasonably warm temperatures, are chinooks in the Great Plains of North America and foehns in Europe.

Chinooks descend the eastern slopes of the Rocky Mountains into the United States and Canada. But instead of being pushed by summer highs over the Rockies, they are drawn by winter lows east of the Rockies. Like katabatic winds, when funneled through narrow canyons, they can reach high speeds. And like Santa Ana winds, the chinooks are warmed by compression. Moreover, they can descend on an area so quickly that local temperatures can rise from near freezing in the morning to above 100°F (36°C) by lunchtime.

Foehns (pronounced "ferns") develop in Europe when mid-latitude cyclones approach the Alps from the southwest. Rotating counterclockwise, the air descends the northern slopes, bringing unseasonably warm winter temperatures to northern Europe.

Monsoons

Many areas of the world have characteristically different weather patterns in winter and summer due to seasonally shifting patterns of high- and low-pressure systems. Nowhere is this more evident than in Asia. Not only is Asia the world's largest continent, but it is also crossed east-west by the world's largest mountain range, the Himalayas. This storied mountain range acts as a barrier that blocks the north-south flow of moisture while it alters upper-level winds.

The result is the monsoon: a pattern of intensely shifting seasonal winds. Indeed, the word monsoon comes from the Arabic "mausim," meaning "season" or "wind shift." Although the term is often misused to refer to heavy rains, it actually means a climatic pattern in which prevailing winds shift, and a dry season annually alternates with a season of heavy precipitation.

Though Asia is home to the world's most dramatic and famous monsoon, it is not the only continent where this weather pattern is found. North America has a monsoon as well, appearing primarily over Mexico, and reaching as far north as Arizona.

ASIAN MONSOON

Two meteorological factors drive the monsoon. The first is the seasonal movement of the subtropical jet stream—an upper-level river of air associated with the Hadley Cell, which can move both north and south of the Himalayan range. The second factor is the summer solar heating of the enormous Asian continental land mass, which creates a strong inland thermal low over the

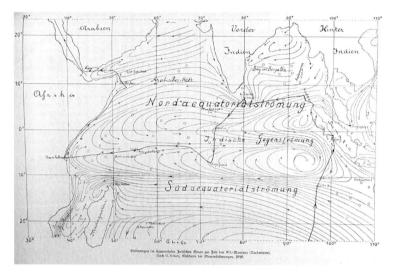

Above: An 1858 drawing of the storm system in the Indian Ocean near the Equator during a monsoon. Top left: Warm Indian Ocean water creates an air mass that eventually is drawn over the hot continent of Asia. The air rises, cools, and condenses, creating torrents of rain known as monsoons.

The Himalayan mountains exert great influence on Asia's weather systems: They block the continent's North-South flow of moisture and contribute to the formation of monsoons.

A monsoon floods the streets in Ahmadabad, India.

Tibetan plateau. The result of these two phenomena is the monsoon.

In the northern winter, winds flow down from the dry Himalayan range in a southwesterly direction toward the Indian Ocean. The descending air compresses and warms, leading to dry conditions over most of India and Southeast Asia. By late spring or early summer, however, solar heating of the Asian continent creates an enormous thermal low that begins drawing moist, unstable air inland from the Indian Ocean. As the ocean air passes over the hot land, it rises and cools, and its moisture condenses into clouds—an effect even more pronounced as it ascends the southern slopes of the Himalayas. That combination releases torrents of rain, sometimes averaging 4 inches (10 cm) per day during peak months.

NORTH AMERICAN MONSOON

The North American monsoon is less dramatic than its Asian cousin, and more poorly understood. It primarily affects the tropical and subtropical areas of Central America and Mexico but its northern edge extends as far north as Arizona and New Mexico. High temperatures over the mountains of Mexico and the western United States play a major role in the monsoon's development and evolution, in a manner similar to what is observed over the Tibetan Plateau. Moreover, as in the case of Asia, shifts in the low-level jet streams bring moisture to the continent from the Gulf of Mexico and the Gulf of California, playing an important role in the daily cycle of precipitation.

After a cool and dry desert winter that rapidly heats to above 100°F (38°C) by May, heavy rain begins falling in early June over southern Mexico. Following the western slopes of the Sierra Madre Occidental Mountains, the rain spreads north. The monthly rainfalls here can be over 12 inches (30 cm), hitting desert areas with tropical rain forest intensity. Equally dramatically, in the autumn after the rains subside and the winds shift, the dry season is truly parched: in Acapulco, for example, rainfall totals about 52 inches (132 cm) from June through October, but only about 3 inches (8 cm) for the rest of the year.

The monsoon rainfall diminishes in intensity toward the north. Rains reach Arizona and New Mexico by July, principally through daily afternoon thunderstorms, often accompanied by lightning and flash flooding.

Rainfall in Acapulco is affected by North American monsoons, and ranges from an average total of 52 inches (132 cm) from June through October, to about 3 inches (8 cm) for the rest of the year.

What's a Forecaster to Do?

The more scientists come to know about weather and climate, the more they recognize the urgency of knowing still more; both money and lives rest on their investigations.

THE WORLD ACCORDING TO GARP

From the late 1960s through the 1970s, a dozen nations joined together their ships, weather satellites, aircraft, ocean buoys, and scientists in the international Global Atmospheric Research Program (GARP). Different investigations conducted simultaneous observations from Equator to poles, from mountains to deserts, systematically monitoring land, sea, and air. Results from GARP, in fact, revealed just how intimate is the interaction between ocean and atmosphere in the global circulation and transfer of energy, moisture, and atmospheric aerosols.

THE THREE PILLARS OF WISDOM

Television meteorologists have long acquainted the public with air masses, warm and cold fronts, high- and low-pressure systems, the jet streams, and even the El Niño–Southern Oscillation. Yet the ability to accurately forecast next week's weather was made possible only through the advent of monumental supercomputers. Especially since the 1990s, computational simulation to model the atmosphere has taken its place as a third pillar alongside observation—both from satellites and the ground—and theory. Together, these three pillars have furthered scientific discovery and continued to refine weather forecasting. The results of such scientific modeling pays off in both economics and lives saved by weather forecasting; meteorologists can now warn of approaching blizzards, hurricanes, or tornadoes from 1 to 72 hours in advance.

There are hard economic reasons to want to make weather forecasts as far in advance as an entire season or even a year. The entire travel industry could benefit from knowing whether next winter would be particularly

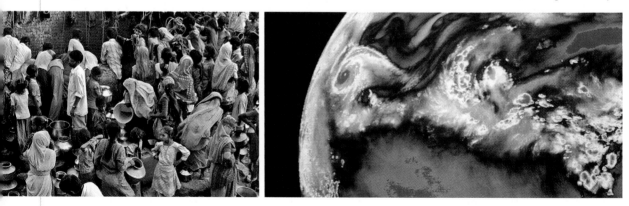

Above left: A village well in Natwarghad, India, draws crowds during the drought of 2003. Above right: This satellite image of Hurricane Floyd was used by the US Air Force Hurricane Hunters to plan their flight through the storm. During this flight the team gathered information such as wind speed, atmospheric pressure, and humidity, which was used to generate predictions about the future course of the storm. Top left: Buoys, like this one being serviced by National Oceanic and Atmospheric Administration's ship KA'IMIMOANA, are instrumented to monitor ocean temperature at varying depths. This information can help forecasters predict El Niño events.

snowy or mild. Citizens in Asia could prepare if they knew next summer's monsoon rains would bring abundant rice crops, or failure and famine. And residents of flood- or hurricane-prone areas could evacuate their homes for safer, higher ground.

Moreover, greater scientific certainty might inspire more deliberate political action. Until recently, there have been long-standing scientific uncertainties and even outright apparent contradictions in the theory, observations, and modeling of climate change. These ambiguities have given intelligent well-intentioned people room for doubt. That doubt has been reduced over the last decade, thanks to increasing scientific analysis of independent sources of evidence, more realistic modeling of the atmosphere, and greater understanding of the behavior of greenhouse gases. Still, remaining scientific uncertainties make some individuals, groups, and nations hesitant to take political and technological action to reduce the emission of greenhouse gases.

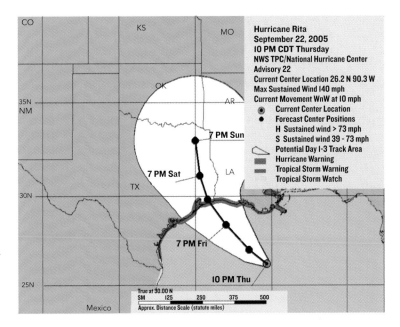

The projected path of Hurricane Rita according to the National Weather Service. Computer-generated forecasts, like this one, alert residents to dangerous weather conditions, providing information that could potentially save lives.

RAIN MAN

Might humans ever control the weather? That question intrigued scientists, science fiction writers, and government agencies alike throughout the twentieth century. Project Stormfury, which was initiated in 1962, was a joint project of the U.S. Department of Commerce and the U.S. Navy, and the first official attempt to answer this tantalizing question. The project's main goal was to try to reduce the destructive power of hurricanes by flying aircraft into them; the aircraft would seed the clouds in the hurricane's eye with tiny crystals of silver iodide. After two decades of intermittent seeding of tropical cyclones in the Pacific and then in the Atlantic, the results were so inconclusive that the program was cancelled in 1983.

Deliberate changes to weather and climate were expressly forbidden by UN General Assembly Resolution 31/72, TIAS 9614 "Convention on the Prohibition of Military or Any Other Hostile Use of Environmental Modification Techniques" (ratified by President Carter in 1979). Nonetheless, the topic is not dead. Half a dozen U.S. western states and one Canadian province have cloud-seeding programs to augment rain and snowfall, reduce hail, and disperse airport fog. Although there are no data to show these programs work, ski areas in Colorado have also experimented with ground-based cloud seeding efforts to produce snowfall. And as recently as October 2004, a private company performed computer simulations for weakening hurricanes; their recommendations focused on early intervention while the systems were still mere tropical storms to discourage their growth, possibly by radiating microwaves at the storms from satellites or by coating the ocean surface with a thin film of biodegradable oil to retard evaporation.

Today, more scientists are concerned about unintentional human effects rather than with deliberate attempts at modifying or controlling weather and climate.

Milestones in Weather

Years bce
4.56 billion years

Earth and solar system are established as the result of a solar nebula.

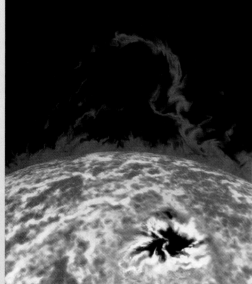

Local concentrations of magnetic fields on the Sun cause sunspots like the dark circle on the lower right, and solar prominences, which in this case appears as a loop-shaped eruption.

3.8–3.4 billion

The first signs of life begin to emerge.

354–290 million

During the Carboniferous period, named for its massive coal deposits, the Earth experiences moist and mild weather and the growth of plant life.

260 million

The Permian period gives rise to reptilian life forms, dry weather, and the formation of the Pangaea landmass.

206–144 million

The Pangaea supercontinent breaks off into two smaller continents, Gondwana and Laurasia. A greener and more humid atmosphere leads to the evolution of more dinosaur species.

65 million

A large asteroid 6 miles (10 km) in diameter apparently strikes Earth, causing massive changes to the Earth's climate, and possibly causing the extinction of the dinosaurs. Known as the K-T extinction, this event marks both the end of the Cretaceous period and the Mesozoic era.

200,000–100,000

Homo sapiens, or modern humans, emerge in Africa.

10,900–9600

Another period of high glacial activity is caused by a halt or reduction of the North Atlantic thermohaline circulation, in what is known as the Younger Dryas event, or the Big Freeze.

6200–5800

Earth experiences the Mini Ice Age, which causes dry temperatures and a worldwide drop in oceanic surface temperatures by about 5.4°F (3°C).

5000–4000

Earth experiences the Climatic Optimum, with the highest global average temperatures for the Holocene epoch.

CE
900–1300

The Little Optimum, or Medieval Warm Period, produces a warmer North Atlantic climate.

1450–1850

The Little Ice Age brings unusually cold climate to parts of Europe and North America around the time of its colonization.

1450

Leone Battista Alberti invents the swinging plate anemometer.

1643

Evangelista Torricelli invents the barometer, and hypothesizes that the air itself has weight in atmospheric pressure.

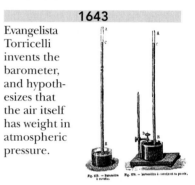

An early diagram of a barometer.

1645–1716

Astronomers note that there is a high scarcity of sunspots. Later research by E. W. Maunder (1851–1928) gave a name to this period, the Maunder Minimum, the coldest period of the Little Ice Age.

1646

Blaise Pascal proves Torricelli is right by testing atmospheric pressure at different altitudes.

1686

Edmund Halley discovers that atmospheric motions are the result of solar heating. He also discovers that there is a definitive connection between barometric pressure and height above sea level.

1724

Daniel Gabriel Fahrenheit invents the Fahrenheit temperature scale, still widely used in the United States.

Swedish astronomer Anders Celsius's best-known contribution to science is the Celsius scale, a temperature measuring system that divides the difference between the boiling and freezing points of water into an even 100 degrees.

1742

Anders Celsius proposes a temperature scale based on equal divisions between the boiling point and the freezing point of water. The Celsius "scale" is still used by scientists worldwide.

1752

Benjamin Franklin conducts his famous kite experiment in Philadelphia, extracting electricity from a thundercloud, and showing that lightning is electrical.

Benjamin Franklin advanced the science of meteorology through experiments with lightning that investigated the nature of electricity.

Franklin had proposed the experiment two years earlier, and it had been tested by François Dalibard of France.

1783

Horace-Bénédict de Saussure invents the hair hygrometer, which measures humidity.

1735

George Hadley (1685–1768) tries to explain the atmospheric phenomenon now known as the Hadley cell, which gives rise to the trade winds.

1803

Luke Howard devises the first classification system for clouds, based on the Linnean method of classification.

1806

The Beaufort Scale is devised by British naval officer Sir Francis Beaufort, as a system for measuring wind intensity based on sea conditions.

1815

The eruption of Mt. Tambora in Indonesia spews a deluge of ash into the atmosphere, affecting the global climate. Sustaining cold temperatures year-round, 1816 becomes known as "the year without a summer."

1859

Robert FitzRoy coins the term "weather forecast."

1835

Gaspard-Gustave Coriolis calculates that the rotation of Earth itself deflects everything from artillery shells to movement of winds away from straight-line travel, in a direction that depends on hemisphere (north or south) and with a magnitude that depends on latitude. This phenomenon becomes known as the Coriolis Effect.

1846

Thomas Romney Robinson invents the rotating cup anemometer.

1856

William Ferrel explains conditions of mid-latitude atmospheric circulation, which later became known as the Ferrel cell.

1856

Joseph Henry formulates the first weather map by collecting weather data by telegraph each day and plotting it on a giant U.S. map hung in the lobby of Smithsonian Institution.

1860

The first weather forecast appears in *London Times*, published by Robert FitzRoy, who a year before coined the term "forecasting the weather."

1863

John Tyndall discovers what will become known as the greenhouse effect.

1883

Indonesian volcano Krakatau erupts, casting ash so high into the stratosphere that sunsets and weather cycles are affected for the next three years.

1891

The U.S. Weather Bureau, which later becomes known as the National Weather Service, is put in full charge of storm and weather predictions by the U.S. Congress.

1900

Wladimir Köppen creates a classification system of Earth's climates.

1902

Léon Teisserenc de Bort suggests that the Earth's atmosphere is divided into two layers, the troposphere and the stratosphere, after conducting over 200 balloon experiments.

1903
Vilhelm Bjerknes begins advocating a computational approach to analyzing weather data with the goal of computing weather forecasts.

1923
Sir Gilbert Thomas Walker develops an explanation of the Southern Oscillation, now linked to El Niño.

A 1920s photograph of the release of a sounding balloon, used to measure atmospheric conditions.

1930
Soviet meteorologist Pavel Molchanov invents the first weather balloon, or radiosonde.

1930s
The U.S. Great Plains experiences the "dust bowl" drought in which parched topsoil is blown around in huge dust storms displacing many farmers already suffering because of the Great Depression.

1931–45
During World War II, the U.S. military discovers that radar can detect precipitation. Radar begins to be used as an tool for studying weather.

1941
Milutin Milankovitch develops the idea that various cycles—such as changes in the ellipticity of Earth's orbit around the Sun—are linked to the occurrence of ice ages.

1946
Cloud seeding, a system of weather modification by interference, is discovered. This method uses silver iodide injected into clouds to induce or delay rain and hurricane events.

mid-1950s
Charles Keeling is the first to raise concerns about global climate change based on his observations of rising CO_2 levels at Mauna Loa Observatory in Hawaii.

1952
Harold Urey proposes that Earth's primordial atmosphere was composed of ammonia, methane, and hydrogen.

1954
Marcel Minnaert publishes his classic book on meteorological optics *The Nature of Light and Color in the Open Air*, reprinted many times since.

1960
The United States launches its first successful weather satellite, TIROS-1.

1969
Herbert Saffir and Bob Simpson devise the Saffir-Simpson Scale for measuring hurricane force and categorizing them based on wind intensity.

1960s–1970s
The Global Atmospheric Research Program and other international cooperation first reveals the intimate connection between the atmosphere and the oceans.

1970s
Solar astronomer John A. Eddy collects and publishes results from Maunder's sunspot research and Douglass's tree-ring data, showing

A dust storm, resulting from drought conditions and poor land-use practices, swirls behind a young farm boy in Cimarron County, Oklahoma, in 1936.

that in past centuries sunspots had virtually disappeared from the Sun in periods that coincided with the Little Ice Age.

1971

Tetsuya Theodore "Ted" Fujita introduces a scale for quantifying the force of tornadoes based on the type of damage they cause.

"Big Joe," the large rock left of center, was photographed during the Viking 1 *mission to Mars.*

1976

The unmanned space missions *Viking 1* and *Viking 2* land on Mars, taking photographs and searching for evidence of life.

Late 1970s

A large hole is discovered in the ozone over Antarctica and theorists conclude that it resulted from the overuse of CFCs and other anthropogenic compounds.

1979

Leslie R. Lemon and Charles A. Doswell theorize that supercell thunderstorms are the weather systems responsible for the formation of tornadoes.

1980

Robert Greenler publishes his classic book on meteorological optics *Rainbows, Halos, and Glories,*

the first to use computer simulations to portray the origins of the phenomena, as well as the first to publish color photos to assist with identification.

1988

The Intergovernmental Panel on Climate Change (IPCC) is established as an independent scientific body to study climate change.

1989

Industrialized nations agree to the Montreal Protocol, which bans use of CFCs in order to prevent further growth of the ozone hole.

1990s

Continued monitoring of the ozone hole reveals an overall pattern as it increases in size through the decades.

1998

A bloc of 169 nations and agencies sign the Kyoto Protocol to the United Nations Framework Convention on Climate Change (UNFCCC), promising to meet standards to reduce CO_2 and other greenhouse gas emissions by 2005. The U.S. is not among the signers.

2005

Hurricanes Katrina and Rita slam into the Gulf Coast of the U.S., causing widespread destruction, death, and the mass migration of people from the region. Katrina is the costliest hurricane to hit the U.S., and the third strongest in U.S. history, while Rita is the fourth strongest.

2006

The hottest year on record for the contiguous U.S. The 25 warmest years on record include each of the past 9 years.

Above: Gujo Fjellanger, Norwegian Environment Minister, signs the United Nations Kyoto Protocol on Climate Change on April 29, 1998. Below: A victim of Hurricane Katrina carries his dog through the floodwaters caused by the United States' most costly weather event ever.

Storm Chasers and Hurricane Hunters

There is something thrilling about getting close to a powerful severe storm—a thrill that may be directly related to the danger of doing so. Storm chasing was popularized by the 1996 movie *Twister*, in which research meteorologists drive perilously close to huge tornadoes to gather field data. While the film had some unrealistic elements, it was largely based on work done by the NOAA National Severe Storms Laboratory in the 1980s.

Such field research, officially known as storm intercept, is an essential part of gathering meteorological data and testing theories. A mobile system, usually a van, can be equipped with Doppler radars, mesonet observation equipment, or even mobile ballooning laboratories. As such, these small weather stations on wheels can observe data and either analyze them on site, or send data back to a fixed laboratory. In recent years, the NSSL has mostly discontinued its storm intercept operations.

Storm chasing now has a widespread and growing base of enthusiasts, ranging from the professional or amateur meteorologist to the curious daredevil. There are even commercial operations that offer storm-chasing excursions,

A tornado categorized as an F4 hurtles toward a storm chaser's car near Manchester, South Dakota.

for a price. It is essential to read up on the practice for safety tips and ethical standards before attempting to hunt down a supercell or a twister; the danger of such violent weather systems is not to be taken lightly.

Left: A portable radar truck, known as a Doppler on Wheels, measures the wind velocity of approaching Hurricane Frances in Florida on September 4, 2004. Far left: A videographer documents Hurricane Gaston as it lands in McClellenville, South Carolina.

Hurricane Hunters

Hurricane Hunters is the name given to pilots who fly into the eye of a hurricane to study the approach and severity of the storm. NOAA operates some planes for this purpose, but the better-known Hurricane Hunters belong to a specialized branch of the 53rd Weather Reconnaissance Squadron of the Air Force Reserve. This squadron of 10 WC-130J planes flies through hurricanes and typhoons, usually on 11-hour flight missions, reporting their findings to the National Hurricane Center. The planes carry dropsondes, which are cylindrical tubes with parachutes attached that carry instruments and radio equipment. The dropsonde measures the air pressure in millibars, essential to computing whether a storm is growing stronger or weaker.

Above: U.S. Air Force Hurricane Hunter pilots in the cockpit of a Hercules plane fly through Hurricane Floyd in September 1999. Each plane makes at least two passes through the eye of the hurricane in order to gather data that will be used to project the course of the storm. Below: The eye of Hurricane Floyd as photographed by the U.S. Air Force Hurricane Hunters team.

Hurricane Hunter planes also measure wind speed, sending data back to the NHS every 30 seconds. Often a plane will pass through the eye of a hurricane four times in one flight.

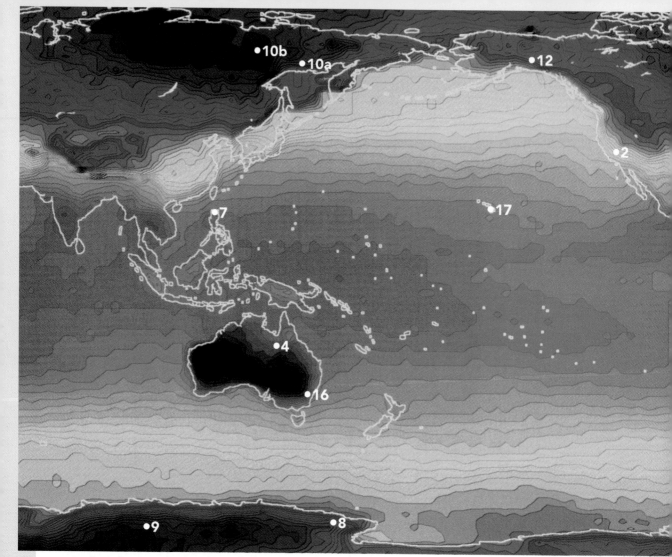

Global Temperature

Highest Temperature Extremes

Locator #	Continent	Highest Temp. °F (°C)	Place	Elevation	Date
1	Africa	136 (57.8)	El Azizia, Libya	367 ft (112m)	Sep. 13, 1922
2	North America	134 (56.7)	Death Valley, Calif. (Greenland Ranch)	-178 ft (54m)	Jul. 10, 1913
3	Asia	129 (53.9)	Tirat Tsvi, Israel	-722 ft (220m)	Jun. 22, 1942
4	Australia	128 (53.3)	Cloncurry, Queensland	622 ft (190m)	Jan. 16, 1889
5	Europe	122 (50)	Seville, Spain	26 ft (8m)	Aug. 4, 1881
6	South America	120 (48.9)	Rivadavia, Argentina	676 ft (8m)	Dec. 11, 1905
7	Oceania	108 (42.2)	Tuguegarao, Philippines	72 ft (22m)	Apr. 29, 1912
8	Antarctica	59 (15)	Vanda Station, Scott Coast	49 (15m)	Jan. 5, 1974

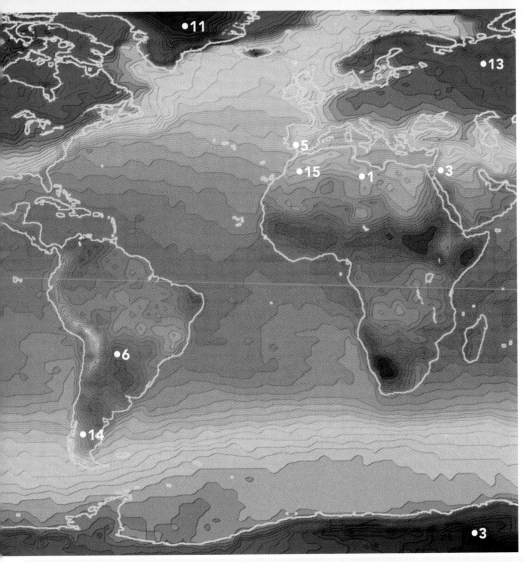

This map of average
surface temperatures
was modeled from
data recorded by
Microwave Sound-
ing Unit (MSU) and
High Resolution
Infrared Sounder
(HIRS) instruments
on a NOAA satel-
lite. Temperatures
are represented
by colors: mauve
(-36°F/-38°C), blue
(-32.8°F/-36°C to
10.4°F/12°C), green
(35.6°F/-10°C to
32°F/0°C), yellow
(35.6°F/2°C to
57.2°F/4°C), pink
and red (60.8°F/
16°C to 93.2°F/
34°C), deep red and
black (96.8°F/36°C
to 104°F/40°C).

Lowest Temperature Extremes

Location #	Continent	Lowest Temp °F (°C)	Place	Elevation	Date
9	Antarctica	-129 (-89.4)	Vostok	11,220 ft (3,420m)	Jul. 21, 1983
10a	Asia	-90 (-67.8)	Oimekon, Russia	2,625 ft (800m)	Feb. 6, 1933
10b	Asia	-90 (-67.8)	Verkhoyansk, Russia	350 ft (106m)	Feb. 7, 1892
11	Greenland	-87 (-66.1)	Northice	7,687 ft (2,343m)	Jan. 9, 1954
12	North America	-81.4 (-63)	Snag, Yukon, Canada	2,120 ft (646m)	Feb. 3, 1947
13	Europe	-67 (-55)	Ust'Shchugor, Russia	279 ft (646m)	January (year unknown)
14	South Amercia	-27 (-32.8)	Sarmiento, Argentina	879 ft (268m)	Jun. 1, 1907
15	Africa	-11 (-23.9)	Ifrane, Morocco	5,364 ft (1,635m)	Feb. 11, 1935
16	Australia	-9.4 (-23)	Charlotte Pass, New South Wales	5,758 ft (1,755m)	Jun. 29, 1994
17	Oceania	12 (-11.1)	Mauna Kea Observatory ,Hawaii	13,773 ft (4,198m)	May 17, 1979

Warming Climate

GREENLAND ICE SHEET 1992

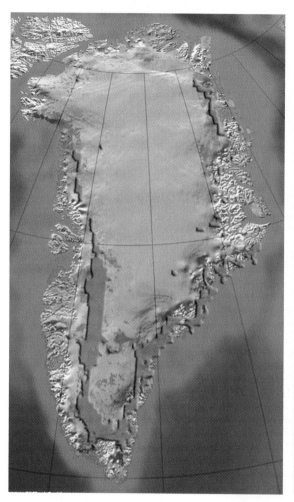

GREENLAND ICE SHEET 2002

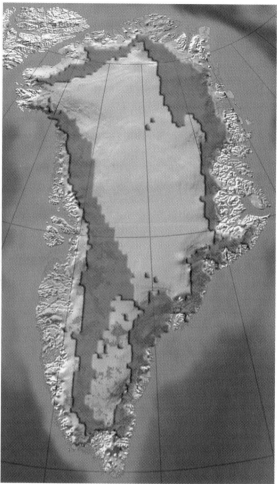

Two graphics released by the Arctic Climate Impact Assessment (ACIA) show the extent of seasonal melting of the Greenland Ice Sheet over a ten-year period, from 1992 to 2002. The ACIA is an international project to study the effects of warming global temperatures on the Arctic region.

The receding ice sheet will mean decreased hunting grounds for polar bears.

ANTARCTIC OZONE HOLE

The ozone hole is an area of Earth with a severe depletion of the layer of ozone. Ozone is a form of oxygen that blocks the Sun's ultraviolet rays, protecting life on Earth. This map, modeled from data recorded by NASA's Ozone Monitoring Instrument, shows the Antarctic region in September 24, 2006. Areas with the least ozone are shown in purple and blue. Green, yellow, and red show regions with more ozone.

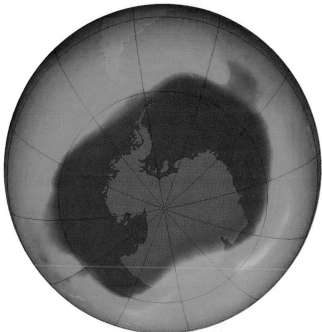

VARIATIONS IN EARTH'S SURFACE TEMPERATURE

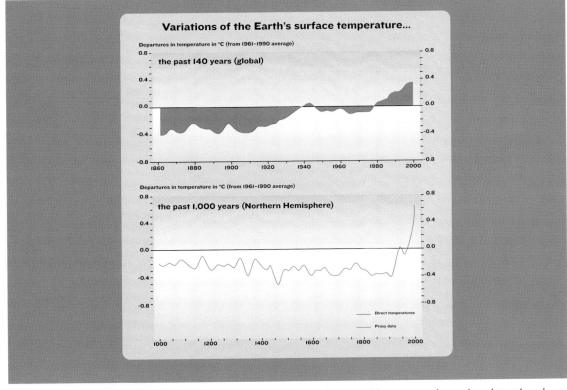

Variations of the Earth's surface temperature...

Departures in temperature in °C (from 1961–1990 average)

the past 140 years (global)

Departures in temperature in °C (from 1961–1990 average)

the past 1,000 years (Northern Hemisphere)

Direct temperatures

Proxy data

These graphs, showing Earth's surface temperatures rising steeply over the past 140 years, were drawn from data released by the Intergovernmental Panel on Climate Change (IPCC). The IPCC was established in 1988 by the World Meteorological Organization and the United Nations Environment Programme to study the problem of potential global climate change. The lower graph acts as a control diagram; while global temperature has fluctuated in the past, the recent sharp rise is a 1,000-year anomaly.

Water Vapor and Sea Temperature

Below: A satellite map shows varying concentrations of water vapor, which makes up clouds, over Earth's oceans. The map is color-coded with the smallest amounts of water vapor shown in light blue and the largest amounts in dark blue. Small yellow areas correspond to areas of high precipitation. Land masses are shown in black with dark green deserts and yellow for ice and snow. This image was taken by an Advanced Microwave Scanning Radiometer, between June 2 and 4, 2002. Right: A computer-modeled map shows sea temperatures based on satellite data taken in July 2001—winter in the southern hemisphere and summer in the northern hemisphere.

Ocean surface temperature is color-coded. The map shows water temperatures ranging from the warm tropical regions to the icy poles. Tropical waters are shown in yellow (95°F/35°C). As water temperature drops, the colors change to red, blue, purple, green, and finally to black for polar waters (28.4°F/-2°C). Earth's land masses are shown in gray.

GLOBAL WATER VAPOR

GLOBAL SEA TEMPERATURE

Measuring Hurricanes

Saffir-Simpson Scale for Hurricane Classification

Strength	Wind Speed (knots)	Wind Speed	Pressure (millibars)
Category 1	64–82 knots	74–85 mph (119–136 km/h)	>980 mbar
Category 2	83–95 knots	96–110 mph (154–177 km/h)	965–979 mbar
Category 3	96–113 knots	111–130 mph (179–209 km/h)	945–964 mbar
Category 4	114–135 knots	131–155 mph (211–249 km/h)	920–944 mbar
Category 5	>135 knots	>155 mph (>249 km/h)	>919 mbar

Tropical Cyclone Classification

Tropical Depression	20–34 knots
Tropical Storm	35–63 knots
Hurricane	64+ kts or 74+ mph

Above: The Saffir-Simpson scale for hurricane classification measures both wind speed and pressure to classify hurricanes. When a hurricane makes landfall, which diminishes its speed, or before it gathers hurricane speed, these weather systems are called tropical storms or tropical depressions. The scale, developed in 1969 by Herbert Saffir and Bob Simpson, charts only storms that form in the Atlantic Ocean and the northern Pacific Ocean, east of the international dateline. Below: This chart shows the progress of Hurricane Katrina from August 23, 2005, when it was still a tropical depression, and August 30, when it was downgraded to a tropical storm as it progressed north over land. Katrina passed over Florida as a Category 1, then gained speed over the Gulf of Mexico, reaching a Category 5 over water, and striking Louisiana on August 29 as a Category 3 hurricane.

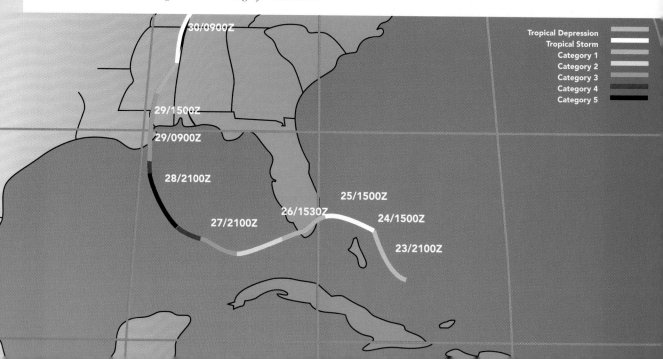

Above: Hurricane Allen, a category 5 hurricane, was the strongest system in the 2005 hurricane season. It achieved wind speeds of 190 mph. Left: A NOAA satellite image of Hurricane Katrina over the coast of Louisiana, taken on August 29, 2005. Below: The most intense Atlantic hurricanes are shown, with maximum recorded wind speeds and central pressure in millibars. This chart is based on information from the U.S. Department of Commerce.

Most Intense Atlantic Hurricanes

Intensity is measured solely by central pressure

Rank	Hurricane	Season	Min. pressure	Max. Wind Speed
1	Wilma	2005	882 mbar	175 mph (282 km/h)
2	Gilbert	1988	888 mbar	185 mph (298 km/h)
3	"Labor Day"	1935	892 mbar	160–185 mph (257–298 km/h)
4	Rita	2005	895 mbar	179 mph (288 km/h)
5	Allen	1980	899 mbar	190 mph (306 km/h)
6	Katrina	2005	902 mbar	175 mph (282 km/h)
7	Camille	1969	905 mbar	190 mph (306 km/h)
8	Mitch	1998	905 mbar	180 mph (290 km/h)
9	Ivan	2004	910 mbar	165 mph (265 km/h)
10	Janet	1955	914 mbar	175 mph (282 km/h)

Measuring Tornadoes

Above and below: Two photographs showing the approach and aftermath of an F4 category tornado near Manchester, South Dakota. Below right: The Fujita scale defines storm intensity by the amount of damage it can inflict on structures in a 3-second gust of wind. The enhanced Fujita Scale of Storm Damage was released in 2006, as an update to the original 1971 F-scale devised by severe storms researcher Tetsuya "Ted" Fujita of the University of Chicago. The enhanced F-scale became operational in the United States on February 1, 2007.

Enhanced Fujita Scale for Tornado Damage

(0-4 For one- and two-family residences)

Number	Damage Indicator	Operational EF Scale
0	Some loss of shingles and siding	65–85 mph (105–137 km/h)
1	Broken glass, uplifted roof, collapse of chimney	86–110 mph (138–177 km/h)
2	Entire house shifts off foundation	111–135 mph (179–217 km/h)
3	Most walls collapse	136–165 mph (219–265 km/h)
4	Total destruction of entire home	166–200 mph (267–322 km/h)
5	Significant damage even to hospitals and government buildings	> 200 mph (>322 km/h)

Rising Sea Levels

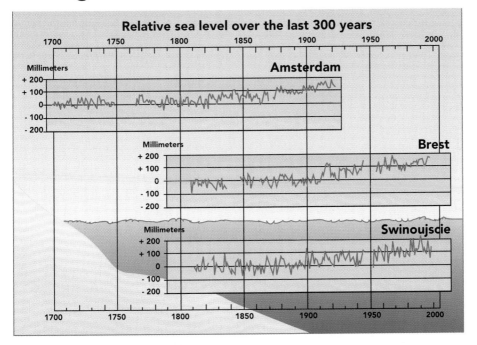

Relative sea level over the last 300 years

Amsterdam

Brest

Swinoujscie

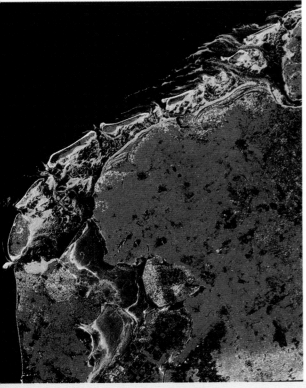

Above: A chart released by the Intergovernmental Panel on Climate Change shows relative sea levels over the last 300 years. Three European locales are used as examples: Amsterdam, the capital of the Netherlands; Brest, a port on the coast of Brittany, France; and the town of Swinoujscie, Poland, spanning the islands of Uznam and Wolin. Left: A satellite photograph shows Dutch land reclamation projects in the Isselmeer, the shallow inland sea once known as the Zuider Zee. The massive dam known as the Afluitsdijk now closes off the Isselmeer from the North Sea. Land is reclaimed by damming off and then draining areas of the Isselmeer to be used for agriculture. Much of the Netherlands is below sea level and must be protected from flooding by a complex system of dunes and dykes. Rising sea levels pose a grave danger to this low-lying nation. Below: Old warehouse buildings rise above the Amsterdam Canal, in Amsterdam, the Netherlands.

CHAPTER 7

CLOUDS

Left: Cumulus clouds, seen here through the spokes of a Ferris wheel. The characteristic "puff ball"-shaped clouds are low-level cumulus commonly seen in fair weather. Top: This satellite image from NASA's Moderate Resolution Imaging Spectroradiometer (MODIS) shows various cloud types over the Great Lakes. The pink-colored portion of the image indicates high-level clouds composed primarily of ice crystals. In the bottom left corner are lower-level clouds such as cumulus or nimbostratus, composed almost entirely of water. The MODIS can distinguish between the various cloud types through readings of the radiant energy emitted or reflected by the clouds. Bottom: Cumulus clouds often dissipate at the end of the day.

Clouds have fascinated poets, painters, scientists, and daydreamers for centuries, and with good reason. In form, clouds range from light, wispy mare's tales to thick, heavy, dark thunderheads. At any moment, clouds cover about half the Earth. Without them we would have no precipitation and perhaps fewer poems.

Clouds are moisture in the air made visible. The clouds' appearance indicates the stability of the air around them. Stratiform clouds, which are layered horizontally, indicate greater stability than cumuliform clouds, which have a lot of updrafts leading to vertical development. Instability is the term used for conditions giving rise to wind, or air moving both horizontally and vertically. So clouds and wind are often related.

All clouds form in the troposphere, especially in the lower half where most atmospheric moisture resides. Indeed, on an average day there may be 40 trillion gallons of water directly overhead in the atmosphere as water vapor, water droplets, and ice crystals. Every day, about 10 percent of that falls to Earth as precipitation. Some of that moisture soaks into the ground, while much of it runs off into rivers, lakes, and oceans. Thus, clouds are an essential part of the water cycle.

In addition to naturally forming clouds, human activities also produce clouds, which exert their own effect on weather and climate.

How Clouds Form

What we call wind is simply the movement of air. Usually when people remark that a day is windy, they are referring to air's horizontal motion. But air also moves vertically—indeed, this vertical motion is what encourages or suppresses the formation of clouds.

WHY CLOUDS FORM

Although the vertical movement of air is usually small-scale compared with the horizontal component, it is nevertheless an important factor in the day-to-day weather.

In a nutshell, rising air cools, allowing its constituent water vapor to condense—forming clouds and perhaps also precipitation; this effect is especially pronounced in the strong updrafts characteristic of thunderstorms. Rising air is associated with low-pressure systems (see chapter 6, "How Weather Works"), which is why lows bring cloudiness and precipitation. In contrast, sinking air warms, causing any condensed moisture to evaporate, thus evaporating clouds and bringing fair weather. Sinking air is associated with high-pressure systems, which is why highs usually bring fair skies and few clouds.

DEW POINT 101

Clouds form when air is cooled to its dew point temperature, usually called simply "the dew point." This is the temperature at which the air becomes saturated with the maximum amount of water vapor that can exist at that temperature, corresponding to a relative humidity of 100 percent. Air cooled to below the dew point is able to condense into liquid dew—hence the name. The dew point is commonly referred to on weather and agricultural forecasts.

The dew point is readily understood from everyday experience:

Imagine it is April, the first spring day when the outdoor temperature has reached 70°F (21°C), sunny, dry and cloudless. The kids want to play in the sprinkler. But moments after getting wet, they rush inside, shivering with cold. Air that feels delightful to dry skin feels uncomfortably chilly to skin that is wet. What is happening? The air has so little moisture (its relative humidity is low) that water evaporates quickly, rushing to fill the void of moisture in the air. This evaporation has a cooling effect. In contrast, imagine it is a humid July morning and already the temperature is a muggy 70°F (21°C). Even at 9 a.m., jumping

Above: A romp through the backyard sprinkler on a 70° day can be a chilly experience if the air is very dry, or has low relative humidity and a low dew point. In such a case, water evaporates quickly, cooling the skin in the process. Top left: Kite streamers indicate the horizontal direction of wind. Vertical wind motion contributes to the formation of clouds.

A view of clouds from an airplane shows several different cloud types; altocumulus clouds, which exhibit some vertical development, appear in the foreground. Such clouds are formed by the condensation of water vapor as it rises in a cool updraft.

air temperature, the air is very dry. Unlike relative humidity, however, the dew point does not change when the air temperature changes, allowing for a direct comparison of moisture content of various air masses.

In another scenario, if the temperature at which air would become saturated is below freezing, it is called the "frost point" instead of the dew point. On still nights when the air temperature drops below the frost point, water vapor directly freezes onto grass as ice crystals (frost) without condensing first into liquid dew, in a process known as deposition.

Once rising warm air cools to below the dew point (or frost point), atmospheric water vapor can condense into tiny water droplets (or ice crystals), forming a cloud.

into the sprinkler doesn't cool anyone off, as wet skin stays wet for 10 minutes or longer. In this instance, the air is nearly saturated (relative humidity is high), so water evaporates only slowly, if at all.

In these two examples, air temperature was identical: 70°F (21°C). On the July morning, the dew point temperature—the temperature at which the air becomes saturated with maximum moisture—was very close to the air temperature. In the April example—characteristic of a dry Mediterranean climate such as that in California—the dew point was tens of degrees lower, possibly under 50°F (10°C).

Dew point is an effective indicator of air's moisture content: When the dew point is high (in the low 70s on the Fahrenheit scale), abundant water vapor is in the air. When the dew point is equal to the air temperature, the air is saturated

and the relative humidity is 100 percent, the dew point never becoming higher than the temperature. Conversely, when the dew point is much lower than the

Early-morning frost on grass occurs when the air temperature drops below the frost point, freezing the water vapor in the air. This process, in which the vapor bypasses the liquid state to become a solid, is called deposition.

Types of Clouds

The first widely accepted classification of types of clouds was that of British pharmacist and amateur meteorologist Luke Howard. Ordering them by shape, Howard proposed in 1802 classifying clouds as they would appear to a ground observer into three categories: cumulus (Latin for "heap" or "pile"), stratus ("layer" or "sheet"), and cirrus ("fiber" or "hair"). To those categories he added nimbus (simply the Latin word for "cloud") as a type of cloud that produced rain.

Years later, Howard and other scientists began to realize the importance of the heights of the bases or bottoms of clouds above Earth. French meteorologist Emilien Renou in 1855 suggested the names altocumulus and altostratus for clouds at middle altitudes. In the 1880s and '90s, two eminent British meteorologists, Hugo Hildebrand Hildebrandsson and Ralph Abercromby,

consolidated the naming of clouds by producing a list of 10 main types based on Howard's scheme, categorized by altitude. They also published a directory of cloud photographs, and formed a cloud committee as part of the 1891 meeting of the International Meteorological Conference. In 1896, at a later conference during the International Year of the Cloud, the naming of clouds and their grouping by altitude was officially adopted, and the *International*

Above: Three types of cumulus clouds are visible in this photograph. The high, wispy clouds in the middle of the frame are cirro-cumulus, the patchy clouds at the top of the frame are mid-level altocumulus, and the heavy rain clouds in the foreground are low-lying cumulonimbus. Top left: High, wispy cirrus clouds over a meadow. Cumulus clouds appear to be piled near the horizon.

Cloud Atlas was launched. The atlas has since been updated several times, the latest version published in 1995.

Today, with minor variations, the classification scheme adopted by the World Meteorological Organization is basically the same, still using combinations of the names originally proposed by Howard. The 10 basic genera recognized today are: high clouds (cirrus, cirrocumulus, and cirrostratus), middle clouds (altocumulus and altostratus), low clouds (nimbostratus, stratocumulus, stratus, cumulus), and clouds that have great vertical extent through all the levels (cumulonimbus). There are also numerous variations and supplemental forms (see sidebar: "Clouds: Genus, Species, and Variety"). Even so, this classification may not be precise, since the heights of cloud bases will vary with local terrain, available average moisture, and weather patterns.

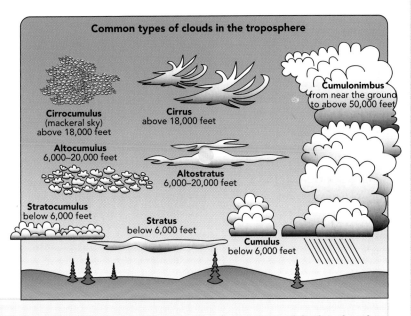

This cloud chart shows the various heights at which different cloud types form. Note that some types of cloud, such as the cumulonimbus, have enough vertical development to inhabit several levels of the troposphere.

CLOUDS: GENUS, SPECIES, AND VARIETY

You look up at the sky and wonder. Is that an altocumulus, you ask, or perhaps a cirrus? It could be either, you realize, and wonder how'd you'd group it if you had to. Exactly this kind of taxonomic perplexity in the biological world led naturalists from Aristotle to Linnaeus to Darwin to categorize animals and plants in a hierarchical structure. They recognized that there may be a subgroup of individuals that has distinct differences from other subgroups, but that still bear a strong family resemblance in physical structure and origin. A domestic dog, a gray wolf, and a fox, for example, all belong to the genus *Canis*. The wolf and the dog are species *Canis lupus,* and the dog is subspecies *Canis lupus familiaris.*

In a similar way the biological world is categorized, the World Meteorological Organization and the American Meteorological Society categorize clouds by genus, species, and variety (analogous to subspecies). The underlying idea is that the genera classify clouds by their main characteristic forms, the species denote peculiarities in shape and differences in internal structure of clouds, and the varieties denote special characteristics of arrangement and transparency of clouds. But even that isn't enough, because there are also categories of supplementary features and "accessory clouds" of minor cloud forms. Moreover, there's a whole category of "mother-clouds" that may give rise to other types of clouds.

The photos in this introductory book show only the 10 cloud main genera: cirrus, cirrocumulus, cirrostratus, altocumulus, altostratus, nimbostratus, stratocumulus, stratus, cumulus, and cumulonimbus, with just a couple of striking variants. For a full look at all 14 cloud species, 9 cloud varieties, and 9 supplementary features and accessory clouds—all with intriguing Latin names and gorgeous forms—consult the photographs in a comprehensive cloud atlas. Hopefully, you will recognize clouds you've seen in the past but couldn't quite place, or you will be inspired to keep an eye out for the many wonderful forms clouds can take.

High and Middle Clouds

High clouds are primarily composed of ice crystals and are usually thin, rarely hiding the Sun. Middle clouds, which can contain both ice crystals and water droplets, may have a similar appearance to the high clouds. Because they are closer to an observer on Earth's surface, though, they appear larger and sometimes can obscure the Sun or Moon. Sometimes middle clouds are associated with light precipitation.

HIGH CLOUDS

High clouds occur above 19,000 feet (about 6,000 m). At such altitudes, where the temperature hovers somewhere around -31°F (-35°C), atmospheric moisture is in the form of fine ice crystals. Over the polar regions, where Earth's surface temperatures are very cold, high clouds made of ice crystals also can be found as low as 10,000 feet (3,000 m).

The most basic type of high cloud is cirrus (abbreviated Ci), usually seen traveling from west to east. Upper-level winds pull these clouds into long wispy hair-like or filamentary streaks with feathery swirls and curls that are sometimes likened to mare's tails. Cirrus rarely thicken enough to obscure the sun or moon, although they can make shadows on Earth look indistinct.

Another type of high cloud is cirrostratus (Cs), which can blanket the sky in an ill-defined sheet or layer of ice crystals that sometimes just turns the sky a hazy, milky blue-white or make the day look slightly overcast. Cirrostratus clouds also produce halos around the Sun and Moon (see chapter 10). In fact, the halos are often the only indication that cirrostratus clouds are present. Often cirrostratus clouds indicate the approach of a storm frontal system, commonly a warm front.

Cirrocumulus (Cc) clouds appear to be a layer of tiny, high-altitude puffs of cotton. Although they can be mistaken for altocumulus (see below), the individual clouds are usually quite tiny as seen from the surface of Earth. Visually striking, they sometimes resemble silvery fish scales, giving rise to the name "mackerel sky." Sometimes they can also appear as waves of clouds that resemble patterns sometimes seen on beach sand—features common to all types of cumuliform clouds.

MIDDLE CLOUDS

Middle clouds occur at altitudes between 6,500 and 19,000 feet (2,000 and 6,000 m), where conditions are warm enough

Above: Cirrostratus clouds, though nearly transparent, can produce a halo effect around the sun. Top left: Cirrus clouds occur at altitudes above 19,000 feet (about 6,000 m) and are formed from ice crystals. The wind at this altitude pulls the clouds into long wispy streaks.

for atmospheric moisture to be liquid droplets, or sometimes a mixture of water droplets and ice crystals.

Altostratus (As) are nearly featureless clouds similar to cirrostratus, except they are lower and thicker as high overcast. They can be so thick that they do not produce halos around the Sun or Moon; instead, those bodies are seen through them as bright spots, perhaps with a white or colored corona surrounding them. Altostratus clouds can diffuse sunlight so much that in the daytime, objects may not cast shadows.

Altocumulus (Ac) clouds are mid-level puffy clouds that can appear either to be patchy, or can occur in distinct linear bands. Some species of alto-cumulus are indicative of particular weather events and have a special significance to weathermen and pilots. For example, a lens-shaped cloud called lenticularis—which sometimes looks like like a stack of pan-cakes—is often found in the lee of mountains where winds flow over rugged terrain.

Altostratus and altocumulus clouds will often occur together.

Above: In this sunset over Lake George, New York, altocumulus clouds create a shimmering, patterned effect known as a mackerel sky. Top: Lenticularis clouds, like this one photographed near Mauna Kea mountain in Hawaii, are so unusual in form that they are sometimes mistaken for UFOs. This unique species of cloud occurs at the crest of a steady updraft of wind near one or more mountains.

Low Clouds and Cumuliform Clouds

Some of the most visually dramatic clouds are closest to Earth, or extend vertically for literally miles. Most often, low clouds are exclusively composed of water droplets, although they may contain ice in colder climates. Low clouds can transform from one type to another. For example, an early-morning deck of stratus overcast can evolve into late-morning stratocumulus, then into early-afternoon cumulus, whose vertical development gives rise to cumulonimbus that produce thundershowers.

LOW CLOUDS

Low clouds range from 6,500 feet (2,000 m) down to the surface of Earth. These dense clouds are usually made up exclusively of water droplets.

Stratus (St), the lowest of the low clouds, is a featureless layer that can cover the entire sky with a blanket of gray overcast without any well-defined outline. The clouds may be only a few hundred yards thick, but the layer can literally extend horizontally over several states. Where the bases of stratus clouds actually rest on the ground, we perceive them as fog. Stratus clouds often appear as a solid deck of overcast, but also can be scattered, although individual cloud elements have ill-defined edges.

Stratocumulus (Sc) clouds have their bases at a uniform altitude and show some vertical development. Viewing the two cloud types from the ground, one can distinguish Stratocumulus from stratus because the former are mottled with dark and light areas, or may look furrowed or have a streaked appearance.

Nimbostratus (Ns) clouds are always associated with a steady drizzle of rain or snow, although it may range from light to heavy. It is very difficult to judge the height of the cloud base because nimbostratus is uniformly gray and ill-defined (the word "dreary" comes to mind). Nimbostratus clouds can range from very low levels up to middle levels— indeed, some meteorologists consider them to be middle clouds.

CLOUDS WITH VERTICAL DEVELOPMENT

Clouds exhibiting great vertical development include most of the cumuliform clouds.

Cumulus (Cu) clouds are the most widely recognized type of cloud, even commonly

Bottom: Nimbostratus clouds provide a bounty of rain to this farmland. Top: Cumulonimbus, or thunderhead clouds, develop vertically into an anvil shape. These clouds portend dramatic weather such as lightning, hail, waterspouts, or tornadoes. Top left: The fluffy, vertical development of a cumulus cloud shows that it is created by a somewhat chaotic upward movement of air.

Stratus clouds, also known as fog, form a blanket over the Golden Gate Bridge in San Francisco, California.

featured in children's drawings. Dense and heavy-looking, the edges of cumulus clouds are well defined and sharp, their appearance resembling popcorn or cauliflower. Fair weather cumulus clouds are the puffy, cottony clouds of spring and summer skies that form at a uniform height above the ground, grow during the hottest part of the day, and dissipate toward sundown. Cumulus clouds are formed when masses of cool air move over terrestrial surfaces that are significantly warmer or colder than the air above, the uneven heating triggering turbulence.

Occasionally cumulus clouds may continue their vertical development until they become thunderheads, technically called cumulonimbus. Cumulonimbus clouds are dramatic, tall, dense clouds whose top may be spread out by upper-level winds into a long plume or anvil shape. They signal thunder, lightning, hail, and perhaps even tornadoes and waterspouts. Although the base of cumulonimbus may be only a few thousand yards high, the top may extend almost as high as cirrus. One striking characteristic is their rapid growth, their shapes changing so fast they almost look as if they are boiling. Strong cumulonimbus clouds occasionally have appendages (officially classified as an accessory feature) looking like pouches under the base of the cloud called "mammatus" because of their resemblance to the mammary glands of mammals. Mammatus clouds indicate downdrafts and instability of the atmosphere, and are associated with heavy thunderstorms.

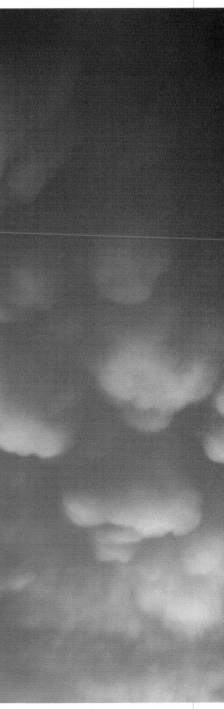

Cumulonimbus mammatus clouds, named for their resemblance to mammary glands, indicate unstable weather conditions and downdrafts.

Artificial Clouds and Human Effects

Not all clouds are natural creations of Earth's atmosphere. Several types, in fact, are inventions unique to human activity just over the last century or so. Formed exceptionally high in the atmosphere, they may be significant for the chemical composition of the atmosphere and climate (see also chapter 12, "Weather, Climate, and Society").

JET CONTRAILS

On some days, the principal clouds visible in the sky are long, white, linear clouds, either parallel or crisscrossing one another. They are frozen vapor from the engine exhaust of jet airplanes flying at cruising altitude in the upper troposphere, where cold temperatures cause the vapor to freeze into ice crystals like cirrus clouds. Called contrails (short for "condensation trails"), under some meteorological conditions, they can persist for minutes or even hours, slowly diffusing and widening into full-fledged cirrus clouds at an altitude of 6 to 8 miles (9 to 12 km). Contrails are so prevalent that they can slightly change surface temperatures by impacting Earth's energy radiation budget.

Another concern is the fact that jet engine exhaust is not solely water vapor. It also includes soot, metal particles, oxidized sulfur, charged molecules, nitrogen oxides, and unburned hydrocarbons. Evidence is strong that these aerosols being injected directly into the upper troposphere and lower stratosphere alter the chemistry of air and influence climate.

POLAR STRATOSPHERIC CLOUDS

Even higher than jet contrails are polar stratospheric clouds. Observed primarily over Antarctica, they have also been seen over Scotland, Scandinavia, and Alaska in northern winter. There appear to be several types. The most spectacular (Type II) are called nacreous clouds (nacreous is Latin for "mother-of-pearl"). Glistening with an iridescence reminiscent of an abalone shell and flowing in a wave-like form, nacreous clouds remain visible for up to two hours after sunset, attesting to the fact that they reside high in the stratosphere at an altitude of some 70,000 to 100,000 feet (20 to 30 km). Less spectacular polar stratospheric clouds (Type I), having no special name, may appear just to be blue or gray.

Despite their eerie beauty, these clouds concern meteorologists. Although the colorful nacreous clouds may be com-

Above: Jet exhaust thins and fans out to resemble naturally formed cirrus clouds. Top left: Contrails from a C-141B Starlifter flying over Antarctica.

Noctilucent clouds, Finland. Noctilucent clouds are the highest clouds in the atmosphere. At an estimated altitude of 250,000 to 300,000 feet (75 to 90 km), they retain light from the sun for hours after it sets.

posed primarily of water ice crystals, NASA aircraft have discovered that the less colorful ones are partly made up of tiny aerosols of nitric acid and sulfuric acid. Their surfaces catalyze chemical reactions that convert rather harmless forms of anthropogenic chlorine (from aerosol cans, older refrigeration units, and manufacturing processes) into an active form (called free radicals) that destroys many ozone molecules in the stratosphere's ozone layer. The result has been a continent-sized hole in the ozone layer that reappears each Antarctic spring and has been growing in size for more than two decades. Thus, the

chemistry humans do on Earth's surface can have profound effects on the upper atmosphere.

NOCTILUCENT CLOUDS

Polar mesospheric clouds, also known as noctilucent clouds (noctilucent is Latin for "night-luminous") are the highest in the sky. First described in the 1880s, they resemble thin silvery or bluish cirrus, sometimes tinged with orange or red, eerily glowing for hours after nightfall against a starry sky. From that fact, meteorologists calculate they are in the upper mesosphere at an altitude of 250,000 to 300,000 feet (75 to 90 km), where temperatures plummet to below -220°F (-140°C).

The composition of noctilucent clouds is unknown. Because they were first described around the time of the mammoth eruption of the Indonesian volcano Krakatau in 1883, at first scientists suspected they might be volcanic dust (perhaps coated with ice) at unusually high altitudes. But instead of decreasing over time, noctilucent clouds are growing bigger, brighter, and more common, in recent years being sighted as far south as Colorado and Utah. In fact, more noctilucent clouds appear in the days immediately following space shuttle launches, suggesting that water vapor and other particulates in the exhaust plume of the space shuttle's main engines may have global effects.

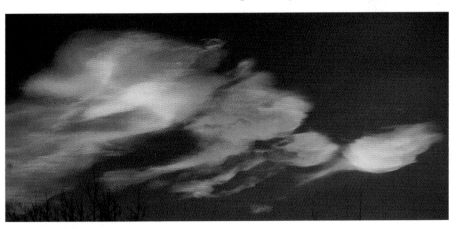

Nacreous clouds, Swedish Lapland. These polar stratospheric clouds are Type II, which means that they are composed primarily of water ice crystals. Type I clouds contain aerosols of nitric and sulfuric acid, and their surfaces catalyze the conversion of chlorine into free radicals that destroy ozone.

CHAPTER 8

PRECIPITATION

Left: A rainstorm on the horizon during a rosy sunset in Four Corners, Utah. Top: Snow weighs down the branches of a poplar tree in Lake Minnewanka, Canada. Bottom: A male bullfrog enjoys a summer shower in Ludlow, Massachusetts. Rain supports all plant and animal life on Earth as well as being an essential aspect of the hydrologic cycle.

Rain, snow, or hail probably first come to mind when people hear the word precipitation, commonly understood as water in some form coming out of the sky. But meteorologists also classify variants of liquid and solid precipitation, such as drizzle or sleet, as separate forms, based on the size of their droplets, the shape and composition of ice, and the nature of their origins.

It might be surprising to learn that dew, frost, and fog are also forms of precipitation—indeed, the principal form in certain climates. What happens to the precipitation both inside and outside a cloud determines whether it ends up as rain, snow, hail, or sometimes even a mixture of types. For example, precipitation may fall directly to the ground; it may be repeatedly caught in updrafts, circulating within the cloud several times before it falls; or it may fall through sub-freezing air. Some precipitation even evaporates before it strikes the ground. Moreover, some forms of precipitation do not even come from clouds.

Precipitation is essential not only to the nourishing of plant and animal life, but also to Earth's overall hydrologic cycle. Rain, snow, and their cousins are responsible for providing most of the freshwater on the planet.

How Precipitation Forms

All precipitation originates when air cools enough for the water vapor within it to condense, or reach the dew point temperature.

CLOUD-FORMED PRECIPITATION

Aside from dew and frost, all other precipitation comes from clouds, within which air rises and cools, allowing the water vapor to condense. Indeed, the rising of air, and its resulting cooling, forms the clouds that subsequently form precipitation. There are at least three different conditions under which clouds holding precipitation can form.

One condition is when air is forced to rise up the slopes of mountains, leading to orographic or relief precipitation on the windward side, and a corresponding "rain shadow" on the leeward side of the mountains. Orographic precipitation is common in island

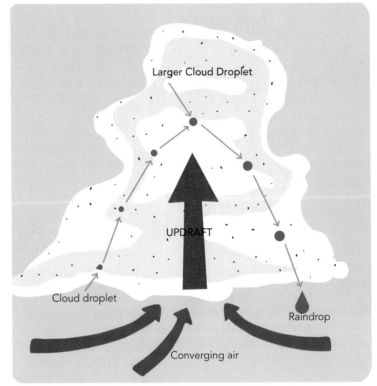

Above: Small cloud droplets are carried by updrafts within the cloud and grow as more water condenses around the nuclei. These drops finally become heavy enough to fall to the ground as rain. Top left: A cloud off the coast of Miscou Island, New Brunswick. Water molecules at the base of this cloud readily evaporate to form the cloud's flat bottom when they come into contact with warmer air close to the Earth's surface.

locations such as Hawaii, where humid ocean air is forced to rise along the steep, high volcanic slopes. It is also characteristic

Ocean air forced up the steep slopes of Poas Volcano in Costa Rica creates clouds that lead to orographic precipitation.

of the western coasts of North, Central, and South America, where thick fog forms on the Pacific slopes of the various coastal mountain ranges.

Another state conducive to the forming of precipitation-bearing clouds is when humid

air is heated by sun-baked land below, forcing it to rise and cool to its dew point. Such convectional precipitation is typical of thunderstorms, like those that form on summer afternoons in tropical, subtropical, and even mid-latitude regions such as Florida and the eastern and midwestern United States.

Precipitation-bearing clouds also originate along warm fronts and cold fronts where the warmer, less-dense air is forced to rise over the colder, denser air. Such a condition leads to frontal precipitation, characteristic of the middle latitudes, especially in the colder months.

WHY CLOUDS DON'T FALL

Not all clouds bring precipitation. That is because cloud droplets (the microscopic droplets that form clouds) are orders of magnitude tinier than water droplets need to be for gravity to make them fall to Earth as rain.

The typical cloud droplet is only about 0.02 mm (0.0008 inch) across. Although cloud droplets do fall because of gravity, they are slowed by air resistance, buffeted by winds, and swirled upward by warm updrafts. Even in still air, when their downward drifting is undisturbed, cloud droplets are so tiny they readily evaporate as they fall from the saturated environment of the cloud; this is the reason fair-weather cumulus clouds have flat bases.

Rain or other precipitation arises only when some droplets, ice pellets, or ice crystals can collide and coalesce with other

Humid air warmed by Florida's land mass rises and cools to its dew point which creates rain in thunderstorms like this one over the Beeline Expressway in Orlando, Florida.

droplets and grow larger, eventually becoming massive enough to fall out of the cloud and reach the ground.

CONDENSATION AND SUPERCOOLING

Water vapor will condense into cloud droplets only in the presence of aerosols (microscopic particles of dust, volcanic ash, salts, or even pollutants), which act as condensation nuclei.

Similarly, ice does not spontaneously form simply because air reaches saturation at a temperature below freezing. Indeed, between 32°F and 25°F (0°C and -4°C)—actually, even as low as 14°F (-10°C)—most clouds are formed not of ice crystals or other forms of ice, but of supercooled water droplets, that is, droplets at temperatures well below freezing but still existing as a liquid. At temperatures anywhere close to freezing, ice crystals or ice pellets will form only in the presence of ice nuclei. Supercooling of water also happens on or near the ground,

but is much rarer because the lower troposphere has so much dust, pollen, and other particulates that act as ice nuclei. Disturbances from wind also cause water to crystallize into ice.

Only at temperatures colder than -22°F (-30°C) are clouds made up primarily of ice crystals. At higher temperatures, they are commonly made of a mixture of ice and supercooled water droplets. Moreover, temperatures have to plunge to a frigid -40°F (-40°C) before water vapor spontaneously condenses into ice crystals in the absence of ice nuclei.

For purposes of understanding precipitation, clouds are classed as warm and cold. Warm clouds are warmer than freezing, so all the moisture is as liquid water droplets of various sizes. Cold clouds are colder than freezing, and their moisture may be supercooled water, ice, or some mixture. Clouds may also be stable stratiform clouds with little vertical motion, or unstable cumuliform clouds with much vertical mixing and updrafts.

Dew and Frost

Dew and frost condense directly from atmospheric water vapor onto cold or moist surfaces, without any processes within clouds. In fact, dew and frost appear on nights that are absolutely clear, without clouds acting as an insulating blanket to trap heat (long-wavelength infrared) near the ground.

DEW

On cloudless nights, surfaces radiate away their heat most efficiently into the black sky, chilling the air immediately above. If the air is still and cools below the dew point, moisture will condense on cold surfaces (such as the metal roof of an automobile). Dew also readily condenses on vegetation, which is moist from the envirotranspiration (essentially the breathing) of its leaves, and whose moisture slightly raises the dew point.

If the dew point temperature is above freezing, water vapor will condense as liquid dew. Dew is also the condensed moisture that forms on the outside of a glass that is holding a cold drink. This indicates that the glass is colder than the dew point of the air in the room.

FROST

If the dew point temperature is below freezing (sometimes called the "frost point temperature"), water vapor from the air will directly condense as ice crystals on vegetation and other surfaces in direct deposition, the reverse of sublimation, without first going through a liquid phase. Like dew, frost—also called hoarfrost—forms only during cloudless nights when the air is calm.

From a distance, frost appears as a white layer on grass and leaves, quickly melting as it is warmed by the rising Sun. Up close, the frost can be seen to be composed of delicate, feathery

Above: A powdery frost clings to the grasses and plants of a field in Grand Teton National Park, Wyoming. Top left: Delicate rays of hoarfrost are formed from the deposition of water vapor directly into ice crystals, the reverse of sublimation.

Moisture, from water vapor, condenses as liquid dew on vegetation as night air cools, creating sparkling gems that quickly evaporate in the morning light.

ice crystals (called hoar crystals) as intricate and beautiful as the branches of snowflakes. Frost is of concern to farmers, because an intense or late frost in spring can kill young vegetation, destroying a crop.

FROZEN DEW

Frozen dew differs from frost in structure and origin. It is ordinary liquid dew that later froze when overnight temperatures plunged below freezing after the dew condensed.

Frozen dew can annoy motorists especially when it forms on the windshield of a car, as the sheet of ice it can form cannot easily be brushed or scraped away until it melts.

A road surface, covered with fine glass beads, produces a halo around the shadow's head, an effect called heiligenschein.

DEW OPTICS

Stand looking at a grassy lawn full of dewdrops as the Sun is rising at your back, casting your long shadow across the grass. Surrounding your head's shadow may be a bright white halo. This phenomenon is called the heiligenschein, for the German words meaning "holy light" or "halo."

The heiligenschein forms because dew drops are spherical; sunlight is internally reflected directly back along the direction from which it came in what is called retroreflection. The halo is especially intense when dew drops form on grass that has fine hairs that suspend the droplets a short distance above the grass blades, allowing them to maintain their spherical shape. The same optical principle makes highway signs, license plates, and some crosswalks—which all have surfaces covered with fine glass beads—highly retroreflective when illuminated by automobile headlights.

Rain

Although all rain originates within clouds, not all rain is alike. Some winter and spring rains are almost like silent mist against which an umbrella is virtually useless, whereas summer thunderstorms open the skies to downpours that slap the pavement and drum the roof. The difference is due to the size of droplets: Smaller, lighter droplets are more affected by air resistance, so they fall more slowly than larger, heavier droplets.

WHERE RAIN COMES FROM

Rain may originate either in warm clouds (where droplets collide and coalesce until they become heavy enough to fall to the ground) or in cold clouds (where frozen or mixed precipitation grows until it is heavy enough to fall, melting into rain en route).

Although cartoon drawings show raindrops as teardrops extended vertically with a point at the top, real raindrops are nearly spherical if they are about 1 mm (0.04 inch) or smaller. Larger ones, however, are deformed by air resistance as they fall, so their bases are flattened. In fact, the largest thunderstorm raindrops are only perhaps half as thick vertically as they are wide horizontally.

STANDARD RAIN

Raindrops are typically between 1 and 2 mm (0.05 and 0.1 inch) in diameter. They can be heard as a light tapping or rustling when they hit the ground or vegetation, and they make visible ripples in puddles. Rain can be produced only by clouds having significant vertical depth, perhaps greater than about 3,000 feet (1,000 m). Although raindrops usually form in warm clouds, they may also originate as snowflakes or ice pellets in cold clouds that melt as they fall through warmer air layers before hitting the ground.

Meteorologists commonly consider light rain to be when individual raindrops are easily visible, and the accumulation of rainfall is less than 0.5 mm (under 0.02 inch) per hour—a condition commonly called "sprinkling."

In a moderate rain, individual raindrops are hard to distinguish,

Top: Heavy rain pours down on a Douglas fir tree. Heavy rain can be composed of droplets as large as a third of an inch across (4 or 5 mm) and can fall as fast as 27 feet (9 meters) per second. Bottom: Raindrops take the shape of flattened spheres as they fall rather than the commonly illustrated teardrop. Top left: Raindrops are typically between .05 and .1 inch (1 and 2 mm) and create visible ripples in puddles.

and rainfall is accumulating up to 4 mm (around 0.2 inch) per hour.

Heavy rain is anything above that, where the rain looks as though it is falling in sheets, and the gutters may be rushing with water. In heavy rains, the droplets are larger—perhaps 4 or 5 mm across (0.16 to 0.20 inch), and fall faster, some 9 meters (27 feet) per second. Thunderstorms may have individual raindrops that are quite large, with drops documented as large as 8 mm (a third of an inch) across, though this size is rare.

Showers are brief rainfalls, regardless of intensity, that may last for minutes or less than an hour, as opposed to steady rains that can persist for one or several days. Showers usually come from small storm systems, especially local summer thunderstorms, whereas steady rains may originate in storm systems that cover large areas.

DRIZZLE

Drizzle is not sprinkling. Drizzle is the falling of droplets much smaller than those characteristic of ordinary rain, only 0.2 to 0.5 mm (less than 0.01 to 0.02 inch) across. Because they are so tiny, drizzle droplets fall slowly—only 3 to 6 feet (1 to 2 m) per second and moisten the ground with hardly a sound or a visible ripple in puddles.

Drizzle is produced primarily by warm stratiform clouds having a fairly shallow vertical depth of less than 1,000 feet (300 m). The clouds also have to be at a relatively low altitude, and the air below them humid, so that the drizzle droplets do not evaporate before they reach the ground.

VIRGA

Virga is the technical name for rain or snow that does not fall to the ground, usually because the individual droplets or snowflakes are so small that they evaporate in mid-air before reaching Earth's surface.

Virga can sometimes be seen beneath lower-level or middle-level clouds, however, as parallel streaks or wisps of light and dark that fade from view some distance below the bases of the clouds.

Rain falls from a cloud but evaporates before it reaches the ground. This form of precipitation is called virga.

Freezing Precipitation

Sleet, freezing rain, and other freezing precipitation can create ice storms dangerous to pedestrians and drivers, and can spell the ruin of a fruit crop if it befalls an orchard of trees setting buds—but can also temporarily transform a winter landscape into a breathtaking, crystalline work of art.

FREEZING DRIZZLE OR RAIN
Freezing drizzle or freezing rain originates as drizzle or rain near freezing temperatures, then falls through much

Rain freezes in patches of black ice, a transparent glaze that creates treacherous driving conditions and often causes accidents like this one on a road near the Bosnian town of Gornji Vakuf.

colder air, becoming supercooled. When the supercooled water reaches the ground, it freezes on contact to form a slick, transparent glaze that simply makes pavement look wet. Because such treacherous ice is virtually invisible to motor vehicle drivers, except for making asphalt look a bit darker, it is sometimes called "black ice."

Left: Ice encases the branches of a bush. Ice storms can severely damage vegetation, weighing tree limbs until they crack and break off. Top left: Supercooled water droplets flow a short distance before freezing and forming these evenly spaced icicles on a fence in the Rocky River Reservation of the Cleveland MetroParks, Ohio.

If conditions are right, the supercooled water droplets can flow a short distance before freezing solid, encasing every twig and bud in transparent ice and creating icicles with regular, even spacing. In major ice storms, the weight of the ice can break tree limbs or pull down utility wires, causing widespread damage.

SLEET

Sleet, more accurately called ice pellets, differs from hail in that it is a cold-weather phenomenon. Ice pellets are formed either by rain that fully freezes on its way down to the ground (unlike freezing rain, which is still liquid when it reaches the surface), or by snowflakes that melt and refreeze on the way down. Like hail (see page 134), they are hard and brittle, clattering and bouncing when they hit the ground, and sometimes also shattering. They usually occur only in brief showers. Unlike freezing rain or freezing drizzle, ice pellets form a bumpy surface that is opaque white in places rather than being a smooth, transparent glaze.

MEASURING PRECIPITATION

Measuring precipitation is trickier than it sounds, especially if the goal is to have some standard means of comparing totals from place to place.

One long-standard technique is a rain gauge, a circular collector with a diameter of 8 inches (20.3 cm) that funnels its contents into a narrow tube calibrated with depths. Automated collectors including a tipping-bucket gauge, in which the weight of a standard volume of water causes a collecting bucket to tip over and empty itself, noting the time so that the rainfall intensity can be monitored throughout a storm. Still, wind, evaporation, and other factors can affect the accuracy of the measurements. Moreover, outside developed countries—and especially over the 70 percent of the planet that is ocean—measurements are few and far between.

To ascertain the amount of precipitation that falls as snow, rods are used to measure the depth of accumulated snow. Then the water equivalent of snow is calculated, usually equating 10 inches (25 cm) of snow to be equivalent to 1 inch (2.5 cm) of water, or a 10:1 ratio. This, too, has inaccuracies, because snow varies greatly in its moisture content, with a ratio of 50:1 sometimes being more accurate for dry Colorado powder versus 4:1 for a slushy Mid-Atlantic snowfall. So other mechanisms supplement depth measurements by automatically recording weight of falling snow; one example is a snow pillow, a type of antifreeze-filled air mattress that registers the pressure exerted by the weight of a snowfall on its surface. Also, a core of snow may be collected and melted to obtain the water content. The intensity of rain, hail, and snow can also be remotely determined by the strength of echoes returned from weather radar. Special instruments have been devised for measuring the amount of precipitation condensed out of fog by trees and other vegetation in cloud forests. One technique is to expose a fine-meshed net to the fog and then compare the weight of the net when wet to the same net dry. Rain gauge readings have been in use for decades.

Top: A historic photograph of a shielded snow gauge in the northwestern United States measures snowfall as a liquid equivalent. Bottom: A tipping-bucket rain gauge. This system of rain measurement allows for the timed monitoring of precipitation during storms.

Snow

Snow is composed of crystals of ice. Although all snow crystals have a hexagonal structure, their specific shapes have marvelous variety, depending on the conditions under which they form.

The two basic categories of ice crystals in clouds are hexagonal plates (in which the hexagonal sides are much larger than the crystal's thickness) and hexagonal pencils (in which the thickness is much larger than the hexagonal sides). Plates result in what most people think of a snowflake, whereas pencils result in snow needles.

All snow crystals start from individual tiny ice nuclei. Depending on conditions, a flake may grow fairly uniformly, ending up as a large plate or needle. Alternatively, as the nuclei swirl within clouds, water vapor freezes onto them through condensation in the same direct-deposition process that creates rime (see page 134). As their edges grow through riming, the snowflakes take on a feathery branching structure of surpassing loveliness.

SHAPES OF SNOW CRYSTALS

Individual glorious six-sided flakes of intricate beauty—the ones that get all the press as the typical winter snowflake—usually fall in particular circumstances, often at the beginning of a heavy snowfall. The air is usually very cold, and therefore quite dry. Individual snowflakes can grow so large—up to a quarter of an inch (4 or 5 mm) across—that their reflective, flat surfaces act as tiny mirrors, glinting by the light of moon or streetlights to make a blanket of snow appear to glitter.

Six-sided snow needles, although commonly smaller, can also grow to impressive lengths of more than a quarter inch (4 or 5 mm).

What appear to be large puffs of snow falling silently to form deep drifts usually turn out to be aggregates of a dozen or so distinct flakes joined at various angles, growing out one from another. Such aggregates of snowflakes form when conditions in clouds are warm enough that the surfaces of individual flakes are coated with a thin film of supercooled water, causing colliding flakes to stick together.

SNOW GRAINS

Snow grains, also called granular snow or snow pellets, are tiny, white, opaque pellets less than 1 mm (0.04 in). Generally floating to the ground in small quantities from low stratus clouds, they are the frozen equivalent of drizzle: ultra-tiny snow crystals or graupel that have not been caught in updrafts or circulated enough to grow larger. Depending on conditions, snow grains can be mixed with graupel or snow needles.

DIAMOND DUST

Precipitating at ground level only in bitterly cold polar temperatures, "diamond dust" consists of tiny hexagonal plates or pencils with no branching; they drift gently downward, making the air sparkle as if with glitter. More commonly, diamond dust forms at the cold middle and high altitudes in the troposphere where air is stable (no updrafts), giving rise to dramatic displays of meteorological optics in bright sunlight (see chapter 11).

Above: Close-up shot of ice-crystal structure of snow. Right: Snow crystals develop into dendrites, needles, prisms, columns, and thin, thick, and sectored plates according to air temperature and saturation. Top left: Snow crystals form a powdery blanket in Monteagle, Tennessee.

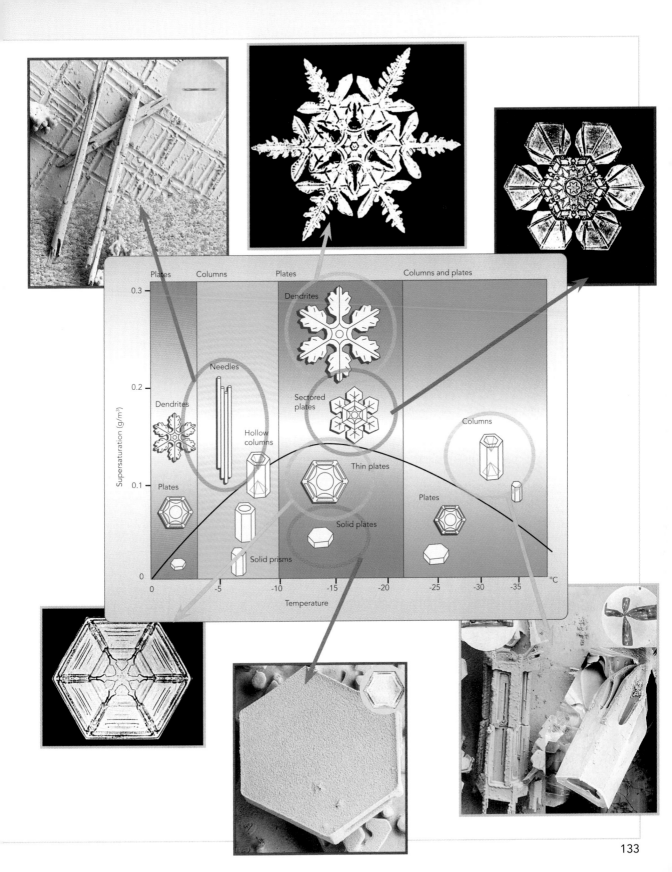

Plates   Columns   Plates   Columns and plates

Dendrites

Needles

Dendrites

Sectored plates

Plates

Hollow columns

Columns

Thin plates

Plates

Solid plates

Solid prisms

Supersaturation (g/m³)

0.3

0.2

0.1

0

0   -5   -10   -15   -20   -25   -30   -35   °C

Temperature

Hail, Rime, and Graupel

Hail (sometimes called "hard hail") and graupel (sometimes called "soft hail" or "snow pellets") form by similar mechanisms. Ice nuclei (in the case of hail) or snowflakes (in the case of graupel) get caught in repeated updrafts and circulate within clouds, colliding with supercooled water droplets that coat them with repeated layers of amorphous (noncrystalline) ice until they take on a semi-spherical shape. Once heavy enough, they fall from the clouds to Earth.

RIME

Rime is ice that looks much like frost but forms by a different

Above: A snowflake dotted with silver drops of rime. Rime condenses onto snowflakes in the clouds after the initial crystal is formed. Top left: Small hail beads among dried grass in the Catskill Mountains in New York.

mechanism, known as accretion. Rime grows when supercooled liquid cloud droplets freeze onto ice crystals, causing them to grow into larger snowflakes, or when supercooled cloud droplets collide with trees or towers on mountaintops, encasing them in opaque white ice.

Riming is the process of supercooled cloud or fog droplets accreting on ice crystals, and can happen even in the presence of wind. The droplets freeze on the crystal, piling up and changing its shape, so that when it falls, it is seen as heavy, icy snow. Rime is what grows on the riggings of sailing ships in icy winter storms, with the crystals themselves growing so they point to windward (that is, in the direction from which the wind is blowing), sometimes becoming inches (centimeters) long.

Riming is also the process by which snowflakes, graupel, and hail grow while being circulated through clouds on updrafts. When enough cloud droplets accrete on a snow crystal, encasing it in rime, it becomes graupel.

GRAUPEL

Graupel, white and opaque, is sometimes also called "snow pellets" or "tapioca snow" for its size and appearance. At casual glance, graupel looks like small

Top and middle: Column and needle ice crystals show bits of rime, which appear like little globs on their surface. Bottom: This ice crystal is so covered with rime that the shape of the initial crystal is unrecognizable.

Hailstones in Leipzig, Germany, show concentric rings that indicate how they are formed. A small ice crystal accumulates layers of ice as warm updrafts and cool downdrafts move it vertically within a cloud.

hail but behaves quite differently. Whereas hail falls straight down and clatters like gravel hitting the ground, graupel is easily swirled by the breeze and is almost noiseless when it bounces from window glass or a car hood, behaving almost like tiny Styrofoam pellets. Lighter and less dense than hailstones of the same size, graupel sometimes has a spongy resilience when squeezed between thumb and forefinger, or is brittle enough to shatter on impact. Its light weight and springiness or brittleness are due to the fact that its structure traps pockets of air, between the frozen rime droplets, instead of being solid ice. Although graupel can get as large as an inch across, it is usually much smaller and seldom does damage.

Graupel can originate high in clouds any time of year, but usually makes it to the ground only during winter months. It forms especially downwind of a relatively warmer body of water over which colder air flows, warms, rises, and circulates, such as occasionally happens around the Great Lakes in the United States. Most lake effect snow is not graupel, but graupel may form in those conditions if the air is fairly warm.

HAIL

If graupel or ice pellets continue to circulate in vigorous updrafts in high summer thunderheads (cumulonimbus), they will collide with supercooled water droplets and accumulate layer after layer of ice, eventually becoming heavy enough to fall as hail. Most commonly, hail accompanies severe summer thunderstorms. Although occasionally falling in all parts of the United States, hail is most common and potentially most hazardous in the Great Plains.

A hailstone sliced in half often reveals a series of concentric opaque and translucent rings, each ring representing a circuit through a cloud; opaque rings indicate where air bubbles are trapped (usually when the hailstone grew quickly through riming), whereas translucent rings indicate where the hailstone grew more slowly by colliding with liquid droplets, which freeze when the stone is carried to higher, colder layers in the cloud.

Most hail is pea-sized, although hailstones may attain the size of marbles, golf balls, or even baseballs. Hailstones larger than marbles can cause serious damage, breaking windows, punching holes through the roofs of houses, killing livestock and people, and flattening whole fields of growing corn within minutes. Indeed, hail destroys a total of 1 percent of the world's crops each year, accounting for $1 billion in damage within the U.S. alone. Rarely, hailstones have been documented to be up to 6 inches (15 cm) across and as heavy as 3.5 pounds (1.6 kg)—virtual cannonballs slamming to Earth at faster than 110 miles per hour (180 km/h).

Just as there are specific locations on Earth most beset by tornadoes, hurricanes, or Nor'easters, there are specific locations whose local topographic and meteorological conditions are most conducive to forming hail. Prime hail country is in the mid-latitudes downwind of a major mountain chain—including northern India and Eastern Colorado, Nebraska, and Wyoming. In the United States, hail is least frequent in Florida or along the Pacific Coast, although it is not unknown.

Fog

Fog is basically a cloud resting on the ground. This fact is clearly evident when one is traveling in the mountains and finds oneself enveloped in fog when reaching the visible base of the clouds. Many times, however, fog is only local and patchy, forming only in pockets or valleys or around ponds or streams. The international definition of fog is fine water droplets, typically the size of cloud droplets, suspended in the atmosphere that reduces visibility to below 0.6 mile (1 km).

Fog can originate in various manners, but always occurs when the ambient temperature reaches the dew point. It can also exist as water droplets when the temperature drops below freezing. In exceptionally cold conditions, fog can actually be composed of ice crystals, a circumstance known as ice fog.

A FOG BY ANY OTHER NAME

Radiation fog occurs when a humid layer near the ground cools until it reaches the dew point, commonly overnight, especially when the air is still or stirred by only light breezes. Radiation fogs are common in coastal areas or over wetlands. In certain climates, such as California's Central Valley and

Top: Cold air moves across the warmer surface of this lake in Darhad Valley, Mongolia, and produces steam fog. Bottom: Humid air from Telegraph Creek in British Columbia, Canada, moves uphill. When the air reaches an altitude where the temperature is below the dew point, it cools and the water molecules form upslope fog. Top left: Advection fog forms a mist over the Yukon River in Canada.

Radiation fog hovers over marsh grass in a lake in Voyageurs National Park, Minnesota.

FOG DRIP, CLOUD DRIP

Heavy fogs and mists frequently collect on the mountains in many parts of the world, ranging from California redwood forests and the Pacific Northwest to cloud forests in Costa Rica, Mexico, the Andes, Hawaii, Kenya, Eastern Europe, and Asia. Moisture from the fog droplets accumulates on the leaves and needles of the vegetation and runs down the stems of plants, or drips directly from leaves onto the soil below. This process has been variously termed "horizontal precipitation" or "occult precipitation."

The amount of moisture captured by forest canopies is so great—annually amounting in some areas to an equivalent of more than 2 feet (0.6 m) of rainfall, or between 40 and 90 percent local annual totals—that the fog water or cloud water is significant in recharging local stores of groundwater. In some remote communities, local residents have even erected fine-meshed nets on hillsides to capture fog moisture for drinking water.

occasionally closer to the coast, an exceptionally dense shallow fog, known locally as tule fog ("tule," pronounced "TOO-lee," being the name of the rushes in the wetlands over which it commonly forms), can reduce horizontal visibility to less than 100 feet (30 m) while still allowing clear views of moon and even stars overhead. Although such a low-lying radiation fog is often called ground fog, strictly speaking according to U.S. weather and aviation practice, ground fog is defined as a fog that obscures less than 0.6 of the sky and does not extend to the base of any clouds above.

Advection fog arises when humid, warm air moves across a much colder surface, such as snow or ice or even cold ocean. As a result, advection fogs are most common in late winter or early spring.

Steam fog, in contrast, occurs when much colder air moves across a warmer surface, such as crisp autumn air flowing across a warmer lake or stream at dawn, or when cold air moves across a heated outdoor swimming pool. Seen commonly in the arctic, its curling appearance makes it sometimes known as "sea smoke."

Upslope fog occurs when humid air rises up along a gentle hillside and cools below the dew point, forming fog above a certain altitude.

Frontal fog can occur either before the passing of a warm front or after the passing of a cold front, when rain falls into cooler, stable air, thereby raising the dew point temperature. It can also occur during the passage of fronts when warm, moist air mixes with cooler air.

Six of these large fog moisture collectors supply 80 people in the village of Danda Bazzar, Nepal, with fresh water.

Water Cycle Warnings

All people, plants, and animals depend on the water cycle for survival. No precipitation—or no clean precipitation—and plants die, crops fail, animals starve, and famine spreads. Yet many human actions and their short- and long-term results, have significant consequences for precipitation and the water cycle.

ACID PRECIPITATION

Acid rain first gained widespread attention in the 1960s, but the name doesn't tell the full story. Anthropogenic, or human-induced pollutants can combine with any form of precipitation to cause acid precipitation of many kinds, including acid snow and acid fog, commonly appearing hundreds of miles distant from the original source.

Principal culprits are sulfur dioxide, unburned hydrocarbons, various nitrogen oxides, and ozone. All these substances are pollutants emitted by burning fossil fuels in motor vehicles, metal smelters, or electric power plants. When such gases or particulates combine with the moisture in clouds, they undergo chemical reactions that turn them into various acids (principally nitric acid and sulfuric acid) whose acidity in precipitation varies somewhere between

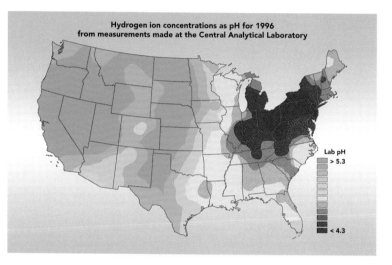

This map shows the varying acidity of precipitation across the United States in 1996. The Northeast has suffered the worst effects of acid rain. Top left: The Sun rises in the polluted skies over Beijing, China. Anthropogenic pollutants combine with water molecules to form acid precipitation.

the acidity of lemon juice and battery acid.

Not only can acid precipitation eat away at building materials, including limestone, glass, steel, plastics, and paint, but it also stunts, sickens, or kills trees and crops, fish in lakes and streams, wildlife, and even has been toxic on occasion to livestock and humans, especially children. In recent years, continental air pollution has created a persistent haze suspended over polar regions in the winter and early spring, especially over the European Arctic. Snow and other precipitation carries haze aerosols down into the snow

pack; during spring thaws, the acidifying pollutants rapidly wash into rivers, briefly raising their acidity to levels toxic to freshwater fish and other species.

DEFORESTATION DECREASING PRECIPITATION

In cloud forests, especially during dry seasons or in the rain shadow of mountains, fog drip can be the sole source of more than 80 or 90 percent of annual moisture, including that returned to the water table. In such ecosystems, cutting down trees to clear land for agriculture or to build cities poses a danger

Left: A boy fishes for brook trout in Big Pine Creek, California. Acid rain has eliminated this species from many of the streams in the Appalachian Mountains. Right: This salmon exhibits a spinal deformity caused by water pollution.

because it removes the very forests that capture the majority of atmospheric moisture and turn it into precipitation.

Some ecologists fear that deforestation could permanently convert cloud forests into semi-arid grasslands or even deserts. Not only could that potentially spell disaster for valleys downstream, whose residents rely on runoff from captured fog water, but such drastic alteration of a local climate might have wide-reaching effects on a region.

GLOBAL WARMING, CHANGING PRECIPITATION

First, a warming planet would accelerate the hydrologic cycle. Warmer air can contain more moisture than colder air. A warmer climate would cause more moisture to evaporate from the oceans, putting more latent heat into the atmosphere. Because water vapor itself is a greenhouse gas, that would further accelerate warming. Although clouds may also cause cooling by reflecting sunlight, the interaction

is complex, and likely to have unknown and mixed results.

Second, a warming planet would change worldwide patterns of precipitation. Several major U.S. and international research reports have calculated potential effects of various degrees of global warming on global patterns of precipitation. The predicted results—already partially observed—are complex and somewhat surprising.

Although overall levels of precipitation are likely to increase worldwide, likely it would be mainly as rain and mainly during heavy downpours during hurricanes and other violent weather events. Mountain snow pack would be less (less snow) and runoff would be earlier in the spring (early warming). Some areas may see a longer growing season, but others (including the U.S. Great Plains as well as Africa) might see longer and more intense droughts.

Drought conditions can be intensified by warming temperatures. Vegetation holds on to moisture and slows evapora-

tion, but as plants die off from drought, leaving bare earth, evaporation increases, further exacerbating the drought cycle. Intensified droughts might lead communities to pump more groundwater. Water levels might lower in the Great Lakes and major rivers, impacting navigation. Water temperatures would also rise; enough warming has already occurred to reduce populations of cold-water fish such as trout and salmon, and also increase the range of tropical fish toward formerly frigid polar waters.

Salmon swimming in a stream in Prince William Sound, Alaska. Acid rain has wiped out the populations of salmon in 14 rivers in Nova Scotia, Canada, and has done similar damage to the salmon population in Norway.

CHAPTER 9

EXTREME, UNUSUAL, AND VIOLENT WEATHER

Every year, thousands of people are killed, injured, or made homeless by severe weather events. Seniors and children die from heat waves or cold snaps; golfers are struck by lightning; houses are shattered into toothpicks by tornadoes and hurricanes, or are swept away bodily by floodwaters. Droughts also take their toll.

While some violent weather is genuinely surprising for its type, timing, or geography—who expects a tornado on Staten Island or in San Francisco?—other weather events are regularly expected with the seasons, such as hurricanes in Florida and the Gulf Coast, or typhoons in the Pacific.

Why is some weather extreme? Actually, violent weather is the norm: Every planet in the solar system with enough of an atmosphere to have any weather at all has truly dramatic or violent weather, ranging from Venus's high-speed winds, to Mars's global dust storms, to hurricanes that rage for months, years, or even centuries on Jupiter, Saturn, Uranus, and Neptune (see chapter 11).

Many large storm systems that bring extreme weather are rotating storm systems, spinning in part by the Coriolis force, and originating from Earth's own rotation.

Left: Multiple cloud-to-ground lightning strikes from a thunderstorm. The top of the thunderhead shows the classic anvil shape associated with these cumulonimbus clouds. Severe storms of this kind can produce heavy rains, hail, and tornadoes. Top: Three men battle 90-mph (144-km/h) winds as they try to escape Hurricane Georges, which struck Key West, Florida on September 25, 1998. Bottom: Farmland threatened by a huge tornado near Manchester, South Dakota, on June 24, 2003, a record day for tornadoes in the state.

Record Temperatures

Simple heat and cold can be extreme, even causing deaths, especially if it is unexpectedly early in a season before people's bodies have become acclimated to seasonal changes.

Biometeorologists—those who study the effects of weather on people—identify two different kinds of temperature extremes in heat or cold waves. The first is called absolute extremes, where temperature may exceed an absolute temperature index in degrees. For example, temperatures of -40°F (-40°C) or 130°F (55°C) would be extremes even for people living in the Arctic or the Sahara.

The second kind of heat or cold waves is called relative extremes, where temperatures exceed some deviation from normal for a specific climate or season, even if such temperatures are routine in another climate or season. For example, 80°F (27°C) is typical for May 20 throughout the southern continental United States, but set a new record-high temperature for Fairbanks, Alaska, in 2002. The relative extreme is usually considered more noteworthy, because people, animals, and vegetation acclimate themselves to the norms of their own specific regions. That 80° day would feel extremely hot for May in Alaska, but would feel relatively normal in Alabama.

RECORD-HIGH TEMPERATURES

Extreme heat can be oppressive to people and animals, especially if combined with high humidity, making it difficult for bodies to be cooled by normal evaporation of perspiration. Vegetation, on the other hand, suffers especially in dry heat, which rapidly sucks the moisture out of leaves.

There is a profound difference in the perceived comfort of "dry" heat versus the comfort of humid heat. Dry conditions prevail when relative humidity is low, such as in arid summers typical of parts of the U.S. southwest and the interior of California, while humid heat is characteristic of summers in the U.S. Midwest and Eastern seaboard. In dry northern California, a summer day that reaches 90°F (32°C) is comfortable beach weather, whereas an overnight temperature of 60°F (16°C) feels chilly enough to need a jacket; in contrast, in New York City where relative humidity is much higher, the same summer

Above: Death Valley, California, where the highest U.S. temperature has been recorded, at 134°F (57°C). Top left: A man leads his camels over sand dunes in the Sahara Desert in Libya. The highest global temperature, 136° F (58° C), was recorded in El Azizia, Libya in 1922.

high would feel unbearably muggy, whereas the low would make evening air delightfully soft for a stroll.

That difference is reflected in the temperature-humidity index charts put out by the U.S. National Weather Service and other agencies.

The highest temperature ever recorded in the United States was 134°F (57°C) in Death Valley, California, just shy of the highest recorded temperature anywhere in the world: 136°F (58°C) in Libya. There are weather stations in Arizona, Kansas, Nevada,

New Mexico, North Dakota, Oklahoma, South Dakota, and Texas that have also recorded 120°F (49°C).

RECORD-LOW TEMPERATURES

The biggest risks to people from extreme cold are hypothermia, when the body's core temperature drops dangerously low, and frostbite, which causes skin and extremities to freeze, beginning with the fingers and toes.

Extreme cold also can kill, especially if combined with high winds, which whip away warmth from bodies and even buildings much faster than still air at the same temperature. To quantify levels of discomfort or actual danger, meteorologists have calculated comprehensive windchill charts, taking into account how air speed accentuates the chilling effects of low temperatures. Even cold that is not extreme by any absolute measure —such as just below freezing—can spell serious consequences to citrus crops or other vegetation in subtropical or tropical climates, or even to people who live in homes without insulation.

The lowest temperature recorded in the United States was -80°F (-62°C) in Alaska, with locations in Colorado, Minnesota,

Slabs of ice in Antarctica. The lowest recorded temperature on Earth was in Antarctica, at -128.6 F (-89.2 C), at the Russian research station at Vostock.

Idaho, Montana, North Dakota, Utah, and Wyoming all having hit at least -60°F (-51°C) themselves. The lowest officially confirmed temperature anywhere on Earth was recorded in 1983 at the Russian research station at Vostok, Antarctica—a cool -128.6°F (-89.2°C). This low may have been exceeded in 1997, when the temperature was unofficially recorded as approaching -132°F (-91°C), a temperature colder than dry ice.

If you like dramatic seasonal temperature changes, the place you want to be is Siberia, far away from any temperature-moderating oceans. At Verkhoyansk, winter temperatures have plunged as low as -94°F (-70°C) whereas summer heat has soared to 98°F (37°C), the greatest annual temperature swing anywhere on the planet.

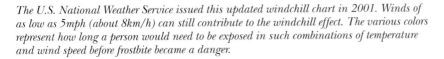

National Weather Service
Windchill Chart

Temperature (°F)

Calm	40	35	30	25	20	15	10	5	0	-5	-10	-15	-20	-25	-30	-35	-40	-45
5	36	31	25	19	13	7	1	-5	-11	-16	-22	-28	-34	-40	-46	-52	-57	-63
10	34	27	21	15	9	3	-4	-10	-16	-22	-28	-35	-41	-47	-53	-59	-66	-72
15	32	25	19	13	6	0	-7	-13	-19	-26	-32	-39	-45	-51	-58	-64	-71	-77
20	30	24	17	11	4	-2	-9	-15	-22	-29	-35	-42	-48	-55	-61	-68	-74	-81
25	29	23	16	9	3	-4	-11	-17	-24	-31	-37	-44	-51	-58	-64	-71	-78	-84
30	28	22	15	8	1	-5	-12	-19	-26	-33	-39	-46	-53	-60	-67	-73	-80	-87
35	28	21	14	7	0	-7	-14	-21	-27	-34	-41	-48	-55	-62	-69	-76	-82	-89
40	27	20	13	6	-1	-8	-15	-22	-29	-36	-43	-50	-57	-64	-71	-78	-84	-91
45	26	19	12	5	-2	-9	-16	-23	-30	-37	-44	-51	-58	-65	-72	-79	-86	-93
50	26	19	12	4	-3	-10	-17	-24	-32	-38	-45	-52	-60	-67	-74	-81	-88	-95
55	25	18	11	4	-3	-11	-18	-25	-32	-39	-46	-54	-61	-68	-75	-82	-89	-97
60	25	17	10	3	-4	-11	-19	-26	-33	-40	-48	-55	-62	-69	-76	-84	-91	-98

Wind (mph)

Frostbite Times   ☐ 30 minutes   ☐ 10 minutes   ☐ 5 minutes

Wind Chill (°F) = 35.74 + 0.6215T - 35.75(V$^{0.16}$) + 0.4275T(V$^{0.16}$)
Where T=Air Temperature (°F)   V=Wind Speed (mph)

The U.S. National Weather Service issued this updated windchill chart in 2001. Winds of as low as 5mph (about 8km/h) can still contribute to the windchill effect. The various colors represent how long a person would need to be exposed in such combinations of temperature and wind speed before frostbite became a danger.

Thunderstorms

Thunderstorms are those that, by definition, have lightning and thunder, and also usually some precipitation, ranging from light to heavy downpours to hail. Although occurring only rarely in the Mediterranean climate of the Pacific coast, which favors a wet winter and a dry summer, thunderstorms are common during summer months in the U.S. Midwest and East Coast, as well as in other parts of the world. Thunderstorms form where conditions favor convection: warm moisture near the surface, lift (so that air rises), and instability (so that rising air once in motion keeps rising).

LIGHTNING

Lightning is the way Earth or any other planet remains electrically neutral. It is essentially a monumental version of the static discharge that zaps a spark from hand to doorknob after you walk across a wool carpet while wearing leather-soled shoes.

In a thunderstorm, droplets of water or chunks of ice or graupel rise and fall in updrafts and downdrafts inside cumulonimbus clouds, becoming electrically charged by friction as they circulate. Positive charges accumulate near the top of the thunderhead; meanwhile, rain and hail carry negative charges down toward the bottom. Because air is a good insulator, the separation of charges builds a gargantuan static charge across the cloud—which discharges as lightning in less than a millionth of a second. Thunder is basically a sonic boom air-pressure wave from the electrical discharge.

Although 90 percent of lightning leaps from cloud to cloud (a hazard for airplanes), 10 percent flashes from cloud to ground, occasionally striking trees, buildings, and other structures on Earth. In fact, the probability of a house getting struck by lightning is surprisingly high, about 1 in 200 over the course of a year, depending on location, which works out to as high as 1 in 8 sometime during its 30-year mortgage. Across the United States, about 50 people a year are killed by lightning.

Above: Forked lightning strikes the ground near an urban area in Finland. Note the anvil shape at the top of the thunderhead. Top left: Both cloud-to-ground (vertical) and cloud-to-cloud (horizontal) lightning can be seen in this photograph, shot among Arizona's saguaro cactus.

Horizontal lightning photographed at Chimney Rock, Colorado. About 90 percent of lightning leaps from cloud to cloud.

SINGLE AND MULTICELL THUNDERSTORMS

All thunderstorms have both updrafts and downdrafts, either of which may be strong or weak. A thunderhead with one updraft and one downdraft is called a cell. A typical single-cell thunderstorm, also called a pulse storm, may pass overhead in only 20 or 30 minutes.

If a storm system has several updrafts and downdrafts traveling together as a unit, the thunderstorm is called multicelled. In a multicell cluster, the most common type of thunderstorm system, the cells are bunched together. Each cell is at a different phase in its life cycle: some nearly spent and weakening and dissipating, while others are at maximum strength, and still others are gathering strength. Thus, multicell cluster storms can last several hours, perhaps producing flash floods if precipitation is heavy.

If the updrafts and downdrafts form a tightly spaced continuous line—sometimes more than 100 miles (160 km) long and moving faster than 60 miles (100k/h)—the storm is called a multicell line or a squall line. Multicell line squalls can create abrupt, strong downdrafts called microbursts (if smaller than 2.5 miles or 4 km across) or macrobursts (if larger), which can be hazardous to aircraft and cause damage to trees and buildings on the ground.

Pulse storms, multicell cluster storms, and multicell line squalls may produce high gusty winds, both soft and hard precipitation, lightning, and weak tornadoes. Nonetheless, none of these three types of thunderstorm is usually considered severe.

SUPERCELL THUNDERSTORMS

The most dangerous type of thunderstorms is a supercell. Much smaller than a multicell thunderstorm, it may be 6 to 10 miles (10 to 15 km) across. Like a single cell storm, it consists of a single updraft, but one that is exceptionally strong with wind speeds of 150 to 175 miles per hour (225 to 260 km/h). Most importantly, the rising winds are rotating in a system known as a mesocyclone.

Supercell thunderstorms, which are rare compared to multicell thunderstorms, are highly organized and can persist for hours. Always considered severe, they can spawn giant hail as large as golf balls or baseballs. They also are the source of strong to violent tornadoes, which sometimes occur in families of half a dozen or more.

A supercell thunderstorm in Kansas sends out an isolated lightning bolt. This storm produced hailstones the size of baseballs.

Twister!

Tornadoes are among the most destructive and dramatic weather phenomena on Earth. Usually spawned by thunderstorms, a tornado is an intense vortex of wind lasting from a few minutes to a few hours.

SUPERCELL TORNADOES

The tornadoes that originate in supercell thunderstorms are without doubt the most dangerous and the most destructive on Earth. The largest ones documented have had wind speeds higher than 250 miles per hour (about 400 km/h), with funnels up to a mile (1.6 km) in diameter, and passing with a noise often compared to the sound of a speeding freight train. The longest-lived tornadoes have lasted for hours and have traveled hundreds of miles, wrecking houses and taking lives all along their paths. The granddaddy of them all was the Tri-State Tornado of March 18, 1925, which lasted a record 3.5 hours and traveled almost 220 miles (300 km) from Missouri across Illinois into Indiana, killing nearly 700 people.

Where swarms of supercell thunderstorms appear, tornadoes occur in outbreaks or families of half a dozen or more at a time. The single largest documented tornado outbreak was the Super Outbreak of April 3–4, 1974,

Above: A small boy gathers his belongings from the rubble left after a tornado destroyed his home in Chak Miran, Pakistan.
Top left: A dark tornado looms on the horizon near Gull Lake, Minnesota.

ranks second, but tornadoes—even killer tornadoes—have struck other nations, including Bangladesh, China, England, Germany, India, and Russia.

Although all 50 U.S. states have had tornadoes, the most famous region for them is in the southern Great Plains, where conditions are most favorable for thunderstorms. Famed "Tornado Alley" stretches north from the Gulf Coastal Plain of south Texas through Kansas into eastern South Dakota. But that area is neither the sole region for tornadoes, nor is it static with the seasons. In the spring, especially March—the most active season for tornadoes—the range extends into the Ohio Valley. In the winter, the primary region for tornadoes is the lesser-known "Dixie Alley" that extends from the Gulf Coastal Plain eastward into Florida. In the summer, a third, unnamed "alley" extends from the Dakotas across Pennsylvania into New York State.

in which 148 twisters killed 315 people from Alabama to Ohio. Sometimes tornadoes occur sequentially, so what may appear to be one long path of destruction might actually have been produced by several tornadoes in a row, the next one picking up where the previous one left off.

Tornadoes may occur during any time of day or night, although like the thunderstorms that spawn them, they are most common in afternoons and evenings when convective conditions are optimal.

FUJITA SCALE OF INTENSITY

In 1971, meteorologist Theodore T. Fujita published a scale by which one could examine the nature of damage from a tornado and infer the intensity of the tornado causing it, most usually rated from weak (F0, blowing down signs) to incredible (F5, blowing houses off foundations). Although Fujita made provision for even stronger classifications, the National Weather Service

does not recognize designations higher than F5. The Fujita scale was reexamined and recalibrated in 2006, in part because the scale of damage does not recognize differences in types of building construction.

In general, tornadoes are roughly categorized as weak, strong, and violent, with the violent ones having winds in excess of 200 miles per hour (300 km/h) capable of leveling homes. Thankfully, only about 2 percent of tornadoes are violent—in fact, most well-constructed homes can sustain a direct hit by a weak or even a strong tornado without undue damage.

TORNADOES WORLDWIDE

The United States is the tornado capital of the world, logging about 800 annually. Australia

"Tornado Alley"

Minnesota
S. Dakota
Iowa
Nebraska
Colorado
Kansas
Oklahoma
Texas

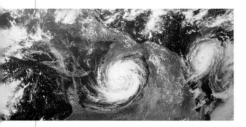

Hurricanes, Typhoons, and Cyclones

First, a vocabulary lesson: Hurricanes and typhoons are the same type of tropical cyclonic storm, but developing in different parts of the world. Hurricane (from a Spanish word) is the name given to tropical cyclones originating in the Atlantic Ocean, the Caribbean Sea, the Gulf of Mexico, and the northern Pacific east of the international date line. Typhoon (from a Chinese word) refers to the same type of storm originating in the northern Pacific west of the international date line.

In other parts of the world, including the southern hemisphere, tropical cyclonic storms are called simply tropical cyclones. In the southern hemisphere, they develop primarily in the southwest Indian Ocean, the Australian region, and the south Pacific. Primarily they affect Australia and southeastern Africa.

ORIGINS

Tropical cyclonic storms are intense low-pressure areas that originate as tropical depressions in the Intertropical Convergence Zone (ITCZ), that persistent region of low pressure and thunderstorms hovering about 10 degrees north or south of the Equator. These storms then intensify into tropical zones and move poleward. Eventually some grow strong enough for

their wind speeds to cause them to be reclassified as full-fledged hurricanes, typhoons, or cyclones.

Tropical cyclonic storms cannot originate right on the Equator itself because the Coriolis force, which increases with latitude, is not strong enough to send them spinning; in addition, they cannot cross the Equator.

HURRICANE FORCE

The intensity of all tropical cyclonic storms is rated according to the Saffir-Simpson scale, with any cyclonic storm of Category 3 or above classified as major. The primary source of energy powering tropical cyclonic storms is the warmth of the ocean water over which they travel. They lose force over land, which is why hurricanes dissipate after coming ashore into the United States. Even so, a hurricane that weakens after coming ashore over Florida can pick up strength again if it continues into the Gulf of Mexico—exactly what happened with infamous Hurricane Katrina in 2005. Tropical cyclonic storms also lose force over cooler waters— under 80°F (27°C)—which is why they rarely hit Hawaii or the California coast, both of which would require a storm following a usual east-to-west track to cross relatively cold water.

Because water warmth is so important, the season for all tropical cyclonic storms is summer and early autumn, after ocean waters have reached their peak summer temperatures. The official northern hurricane season runs from June 1 to November 30, with peak months being August into October. The typhoon season has no official bounds, but usually runs from

Saffir-Simpson Hurricane Scale		
Category Wind speed Storm surge		
	mph (km/h)	ft (m)
1	74-95 (119-153)	4-5 (1.2-1.5)
2	96-110 (154-177)	6-8 (1.8-2.4)
3	111-130 (178-209)	9-12 (2.7-3.7)
4	131-155 (210-249)	13-18 (4.0-5.5)
5	156 (250)	>18(>5.5)

Additional Classifications

Tropical storm	39-73 (63-117)	0-3 (0-0.9)
Tropical Depression	0-38 (0-62)	0 (0)

Above: The Saffir-Simpson Hurricane Scale classifies the severity of tropical cyclones in the Western Hemisphere. Top left: Hurricane Allen swirls over the Gulf of Mexico with another un-identified storm brewing to its right on August 8, 1980.

May through November. Because the seasons are reversed in the southern hemisphere, the peak season for southern-hemisphere cyclones runs from late October through May.

Meteorologists long believed that cyclonic storms could not intensify to hurricane force in the southern Atlantic Ocean because of cooler ocean temperatures, the abrupt pattern of winds, and an absence of typical ITCZ weather. For decades this was observed to be the case. But in March 2004, a southern-hemisphere tropical depression developed in waters that, for the southern Atlantic, were unusually warm—75°F to 79°F (24°C to 26°C). Called Cyclone Catarina, it became a full-fledged Category 2 hurricane with sustained winds of 100 miles per hour (160 km/h), shocking both meteorologists and residents when it struck the coast of Brazil the night of March 27, killing three people.

The Galveston hurricane of September 1900, one of the worst natural disasters in U.S. history, flattened many of this buildings in this Texan city and took between 6,000 and 8,000 lives.

DEADLIEST, COSTLIEST

For hurricanes that have struck the United States between 1851 and 2004, by far and away the deadliest was the one that came ashore at Galveston, Texas, in September 1900, claiming more than 8,000 lives. Hurricane Katrina on August 29, 2005, which claimed at least 1,330 lives, holds the record for being the most damaging hurricane. Katrina also exceeded Hurricane Andrew (August 1992) for being the most expensive ($96 billion versus $33 billion).

The sea temperature at the time of Hurricane Rita is shown with areas in orange measuring over 82.4°F (28°C), a few degrees higher than the temperature needed to form and sustain hurricanes.

Floods and Blizzards

In the United States, floods rank as the twentieth century's number-one natural disaster in terms of deaths and property damage. Floods occur when more water accumulates in an area over a period of time than can drain from the same area in the same time. While the main cause is often excessive rainfall, contributing factors include the saturation of the soil and the blocking of normal drainage outlets. Floods may occur at great distances from the precipitation, and their damage may also be exacerbated by human actions, such as building levees or dams.

TYPES OF FLOODS

Often weather reports refer to minor and major flooding. Minor flooding, which may or may not flow over banks of a stream or river, is usually confined to the waterway's normal flood plain, with the floodwaters usually fairly shallow and slow-moving. Major flooding, on the other hand, may breach levees, dikes, dams, or other structures that are designed to hold back floodwaters, and may inundate city streets and enter houses and other structures. Water is deep and may have a substantial or damaging current.

Although most floods develop over a fairly long time of 12 to 24 hours or more, some develop

Significant Floods of the Twentieth Century

Flood Type	Date	Area or Stream Flooded	Reported Deaths	Approximate Cost (uninflated)
Regional floods	Mar–Apr 1913	Ohio, statewide	467	$143M
	Apr–May 1927	Mississippi River, MO to LA	unknown	$230M
	Mar 1936	New England	150+	$300M
	July 1951	Kansas and Neosho River Basins	15	$800M
	Dec 1964– Jan 1965	Pacific Northwest	47	$430M
	June 1965	South Platte and Arkansas River, CO	24	$570M
	June 1972	Northeastern United States	117	$3.28B
	April 1983– June 1986	Shoreline of Great Salt Lake, UT	unknown	$621M
	May 1983	Central and NE MI	1	$500M
	Nov 1985	Shenandoah, James, Roanoke Rivers, VA and WV	69	$1.25B
	Apr 1990	Trinity, AR, and Red Rivers in TX, AR, and OK	17	$1B
	Jan 1993	Gila, Salt, and Santa Cruz Rivers in AZ	unknown	$400M
	May–Sept 1993	Mississippi River Basin in central US	48	$20B
	May 1995	South-central US	32	$5-5B
	Jan–Mar 1995	CA	27	$3B
	Feb 1996	Pacific NW and western MT	9	$1B
	Dec 1996- Jan 1997	Pacific NW and MT	36	$2-3B
	mar 1997	Ohio River and tributaries	50+	$500M
	Apr–May 1997	Red River of the North in ND and MN	8	$2B
	Sept 1999	Eastern NC	42	$6B
Flash floods	June 14,1903	Willow Creek in OR	225	unknown
	June 9–10, 1972	Rapid City, SD	237	$160M
	July 31, 1976	Big Thompson and Cachela Poudre Rivers, CO	144	$39M
	July 19–20, 1977	Conemaugh River, PA	78	$300M
Storm-surge floods	Sept 1900	Galveston, TX	6,000+	unknown
	Sept 1938	NE US	494	$306M
	Aug 1969	Gulf Coast, MS and LA	259	$1.4B
Dam-failure floods	Feb 2, 1972	Buffalo Creek in WV	125	$60M
	June 5, 1976	Teton River in ID	11	$400M
	Nov 8, 1977	Toccoa Creek in GA	39	$2.8M

Above: This chart shows selected significant floods of the twentieth century, their causes, and locations. Top left: Residents leave their homes in Carlisle, England, due to severe flooding in January 2005. One month's rain fell in 36 hours.

very quickly, in under 6 hours. Such fast-rising floods are known as flash floods, and are most common in hilly or mountainous terrain where water in streams and tributaries flows very fast and accumulates rapidly, and where storms are short-lived but intense. Flash floods are common in arid and semi-arid areas, such as the deserts of the U.S. southwest. An unwitting camper in such a region may set up a tent sheltered from the sandy wind in an arroyo (dry streambed), fall asleep under the stars, and awake to a rush of water from a flash flood caused by thunderstorms in mountains that are miles away.

Snow-melt floods in late winter or early spring can be significant, because snow pack can hold a great volume of water above ground. Often snow melts fairly rapidly during spring thaws, releasing much of the water at once. Also in winter, ice and fallen tree branches and other debris can jam up streams; such ice dams, as they are called, block normal stream flow and force water to spread out onto adjacent land. One danger to humans from either type of flood is that the water released is at near-freezing temperatures.

Along coastlines in regions where hurricanes are common, high winds push the ocean shoreward. Such storm surges, as they are called, drive ocean water inland, a condition exacerbated by heavy rainfall that simultaneously raises river levels. Storm surges are especially common on the east coast of India, and also contributed to the widespread damage caused to the U.S. gulf coast by Hurricane Katrina in 2005.

Regional floods rank among the most widespread of natural disasters, sometimes overspreading thousands of square miles over more than a dozen states. In the United States, regional floods commonly occur in winter with unusually heavy rainfall, when the ground cannot absorb runoff either because it is frozen or saturated from long periods of excessive rain. Major regional floods in the twentieth century that engulfed parts of the Midwest and much of the Mississippi River include those of 1912, 1913, 1927, 1937, and 1993. In other years, lesser floods have still caused a great deal of damage elsewhere in the country.

BLIZZARDS

The U.S. National Weather Service defines a blizzard as having winds of more than 35 miles per hour (about 55 km/h), and enough blowing snow in the air that visibility is less than a quarter mile (400 m). Some blizzards drop several feet of heavy snow on the ground, effectively immobilizing all transportation; others consist primarily of ground-level high winds that lower visibility by blowing already-fallen snow with virtually no additional precipitation. The principal danger posed by blizzards is their arctic wind-chill factors, which can freeze exposed skin in a matter of minutes. In the United States, blizzards are most common across the northern Great Plains, although they also occur in the Midwest, Mid-Atlantic, and East.

Above: A battered awning sags and New York City streets are piled high with snow after the blizzard of 1888. Top: Storm surges, caused by a hurricane in September 1947, drive this huge wave to strike a seawall just north of Miami Beach, Florida.

Droughts

Above: Dead fish, on the shores of the greatly receded Lake Rei, litter the banks after Brazil's month long drought in 2005. Top: Poor farming practices and drought led to dust storms like this one in Kansas, 1935. Top left: An Afghan farmer rakes up the dead remains of his vineyard in March of 2002 after the area's third straight year of drought.

Droughts are a prolonged absence of water, relative to the norms of local and regional climates. A desert or other arid climate is not drought-stricken just because it is the climatic norm for it to receive far less annual precipitation than a forest. Indeed, deserts and arid climates can also suffer droughts relative to their own climatic norms. Moreover, it is also a climatic norm for Mediterranean and other climates with distinct wet seasons and dry seasons to undergo annual seasonal droughts.

Where a drought is a significant and prolonged departure from a climatic norm, however, it commonly develops much more slowly than other types of extreme weather events such as floods and blizzards. Moreover, the effects of a drought may be severe and long lasting; such were the five-year drought that afflicted the northern Sahara Desert from 1968–73, and the "dust bowl" years of the 1930s in the U.S. Great Plains.

TYPES OF DROUGHTS

Droughts are classified in several ways. The two most common measures are the amount of water compared to seasonal and regional norms, and the effects of this lack.

A meteorological drought is one in which precipitation totals (rain and snow) are significantly below the norm for the affected climate. What is considered "significant" is often up to local policy-makers, but it is commonly seen as some fraction, like 50 or 75 percent, over a stated period of weeks, months, or seasons. A meteorological drought has only to do with amount of water from the sky, not the amount of available water in reservoirs or underground aquifers. Usually a meteorological drought is the first sign of drought.

A hydrological drought is one in which the volume of water in a region is significantly below the norms for the time of year, measured in river heights, rate of stream flow, lake and reservoir levels, wells, and other watershed indicators. Hydrological conditions usually lag meteorological conditions, because the water contained in them is usually also from previous years' precipitation, and may be enough to carry a region through a relatively short dry spell. Similarly, hydrological conditions also lag in recovery, because it takes time for lakes and aquifers to recharge after prolonged absence of rain.

An agricultural drought is one in which certain crops cannot grow because of insufficient available water. In areas where crops are not irrigated, a shortage in rainfall will show up in the same growing season; if crops are irrigated, an agricultural drought may not become evident until the next growing season when deep groundwater for irrigation becomes in short supply.

A socioeconomic drought is one in which the shortage of water is so severe or prolonged that it affects human living conditions or livelihoods. Even early in a meteorological drought, people may be required to observe water-conservation restrictions on the watering of lawns and taking of showers. But deep into a hydrological drought, farmers may be wholly unable to raise crops, livestock, or fish, and—in areas dependent on hydroelectric power—even energy supplies may be curtailed.

REAPING WHAT IS SOWN

Natural droughts can be exacerbated by human actions. The dust bowl of the U.S. Great Plains is a true object lesson. Many farmers were first lured to the area by inaccurate reports of its fertility, in part because the first settlers in the nineteenth century arrived during an unusually wet period.

They used agricultural techniques more suited for the humid climate of the eastern states. Although methods of soil conservation were known, they were abandoned in the 1920s and early 1930s because a combination of falling wheat prices and expensive high-capacity plowing technology forced farmers to cultivate large acreage of marginal lands in order to earn adequate income.

When four droughts then occurred in rapid succession throughout the 1930s, farmland became so parched that soil literally blew away in enormous dust storms. Unable to pay their debts, many farm families were forced off their property and migrated to other parts of the country. Poor land-use practices in the Great Plains magnified the agricultural and socioeconomic effects of the droughts and deepened the nationwide Great Depression.

A boy, shown with his cows in southern Ethiopia, was among the region's 16 million threatened by the prolonged drought of 2000.

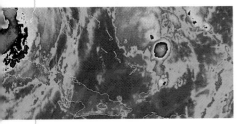

Climate Change: Severe Weather Warning?

For several decades, a number of meteorologists and climatologists have predicted that global warming would lead to greater incidence of severe or extreme weather on Earth. According to the World Meteorological Organization, the National Center for Atmospheric Research, the Intergovernmental Panel on Climate Change, and other bodies, hurricanes could become more frequent and destructive, heat waves and droughts could become longer and worse, and thunderstorms, tornadoes, rainfall, and floods could intensify.

The record activity of the 2005 Atlantic hurricane season made people wonder whether such predictions were being fulfilled. Out of an unprecedented 28 tropical and subtropical depressions, 15 strengthened into full-fledged hurricanes. For the first time, the naming of the storms progressed into the Greek alphabet. For the second time on record, one tropical storm—Zeta—occurred so late it even was active into January of the following year (2006). Hurricane Wilma was the most intense Atlantic hurricane on record, its central pressure (882 millibars) reaching a record low for an Atlantic hurricane and its one-minute sustained winds peaking at 185 miles per hour (295 km/h). And there were the one-two punches to New Orleans delivered by hurricanes Katrina and Rita.

There are many difficulties with assuming that the remarkable hurricane activity of one year fulfills the predictions of extremes, because hurricane seasons routinely vary in intensity and are affected by many complex factors both known and still unknown. Indeed, the 2006 hurricane season, in contrast, was unusually quiescent, with no full-fledged hurricanes even making landfall in the United States, although climatologists are watching carefully.

Still, changes in seemingly more mundane and routine meteorological quantities—in high and low temperatures,

Left: A vulture perches on an anthill and surveys the carcasses of cattle that died as a result of a four-year drought in Ethiopia. This photo was taken on April 13, 2000. Top left: An infrared satellite image shows Hurricane Wilma (left) and tropical storm Alpha (right) near Florida's coast on October 22, 2005. Alpha was the season's 22nd named storm and marked the first time in 60 years of naming storms that the list of alphabetical names for the year was exhausted and meteorologists moved on to the Greek alphabet.

difference between highest and lowest temperatures in a year, warmth of daily lows, number of days with frost, amount and intensity of rainfall, length of wet periods and dry periods, start and length of the growing season—in many regions worldwide, these figures tell their own story. According to a report published in late 2006 by the University Corporation for Atmospheric Research (UCAR), changing patterns in those quantities over past decades are all consistent with a pattern of an overall warmer global climate.

Global Warming Predictions

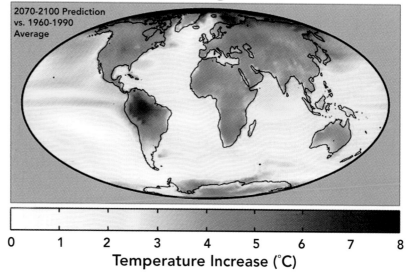

2070-2100 Prediction
vs. 1960-1990
Average

Temperature Increase (°C)

0　1　2　3　4　5　6　7　8

Above: This chart shows the increase in average global temperature due to global warming as predicted by the Hadley Centre's HadCM3 climate model. Below: The National Climatic Data Center's map of Billion Dollar Weather Disasters shows the extreme weather events during 1980–2005 and the damage in billions of dollars that they caused.

Billion Dollar Weather Disasters 1980-2005

1996 / $1.2
2000 / $2.1
1997 / $4.1
1998 / $1.5
1982-1983 / $2.2
1996-1997 / $3.4
1992 / $2.0
1995 / $3.6
1998 / $1.7
1991 / $2.1
1991 / $3.5
1994 / $1.2
1989 / $1.5
1993 / $26.7
1996 / $3.0
1999 / $1.1
1998 / $1.1
2002 / $2.0
2001 / $1.9
1988 / $61.6
2002 / $10.0
2005 / $ 1.0
1999 / $1.4
2003 / $3.4
1993 / $7.0
2003 / $5.0
1996 / $5.8
1999 / $6.5
1993 / $1.3
2003 / $2.5
1999 / $1.7
1994 / $3.7
1998 / $1.1
1989 / $13.9
1980 / $48.4
1995 / $6.8
1997 / $1.1
1986 / $2.3
1998 / $8.3
1993 / $1.3
2005 / $2.0
2003 / $1.6
1982-83 / $2.2
1990 / $1.4
1994 / $1.2
1998 / $1.1
1994 / $1.2
2000 / $4.2
1995-1996 / $6.0
2001 / $5.1
1995 / $3.6
1983 / $4.0
1985 / $2.2
2004 / $7.0
2004 / $9.0
1992 / $2.4
1985 / $2.4
1983 / $5.9
1985 / $2.8
2004 / $14.0
2004 / $15.0
1992 / $35.6
2005 / $16.0
2005 / $125.0
2005 / $16.0
1998 / $6.5
1995 / $2.5
(U.S. Virgin Island)

Legend
⬩ Hurricane
◊ Tropical Storm
▨ Flood
▼ Severe Weather
✳ Blizzard
🔥 Fires
L Nor'easter
Ice Storm
Heat Wave/drought
Freeze

Dollar amounts shown are approximate damages/costs in $ billions.

CHAPTER 10

ATMOSPHERIC DISPLAYS

Left: Rainbows are perhaps the most familiar class of atmospheric optical displays. They form when sunlight passes into raindrops and is both reflected and refracted. In the latter process, the raindrop acts as a prism, splitting the Sun's white light into its constituent colors, but not all rainbows are multicolored, as seen by this red rainbow. Top: A halo around the Sun, photographed near an icebreaker, where a weather balloon is being released. Halos arise when light is refracted and reflected by tiny ice crystals in high-altitude clouds. Bottom: A green aurora over a northern sky. These spectacular displays are caused by charged particles from the solar wind colliding with the Earth's upper atmosphere.

Although the nighttime sky is famed for the Moon and stars, the daytime sky has its own dramatic effects: rainbows, a halo around the Sun, a mirage. A sun pillar—a red or golden pillar of light—stands straight up above the setting or rising Sun. Crepuscular rays—long beams of Sun and shadow—reach down or outward from clouds in the sky.

While such effects have a reputation for being rare, this reputation stems from how rarely they are seen, rather than how rarely they appear—a commentary on how rarely one actually gazes upward at the daytime sky. Look up frequently throughout the day when both Sun and clouds share the sky, and within a month one of some eerie and beautiful optical effects will fall into view.

Atmospheric or meteorological optics is the name for the field of study of such displays. Although they do not cause weather, certain classes of optical displays are more frequent with specific types of weather conditions. One last type of dramatic atmospheric display is visible only at night: the spectacular flickering aurora, produced when charged subatomic particles from the Sun bombard the outermost layers of the atmosphere. Turn the page to view more of nature's own light show.

Atmospheric Optics 101

What happens when a ray of sunlight passes through the atmosphere? There are at least four possible answers, which will explain some of the principles at work in atmospheric optics. First, the ray of sunlight can be reflected from the internal or external surfaces of water droplets or ice crystals, as from a mirror. Second, it can be refracted or bent by traveling through water or ice, perhaps even spreading into its component rainbow colors, as if it had traveled through a prism. Third, it can be

scattered by dust motes in all different directions. And fourth, it can be diffracted, or spread out, when it grazes the edge of particulates smaller than its own wavelength of light.

Two or more of these processes often affect the same ray. Certain meteorological circumstances give rise to particular optical processes. Understanding these correlations can be helpful in spotting sky conditions likely to create breathtaking optical displays.

REFLECTION

Reflection, familiar to anyone looking into a mirror, happens when a ray of light is turned back from the surface of a water droplet or ice crystal. The amount of light the droplet or crystal reflects depends on the smoothness of its surface and the angle of the incoming ray of light.

Smooth, mirror-like surfaces reflect more of the light in one particular direction whereas rough surfaces will diffuse it in all directions. Ice crystals usually have

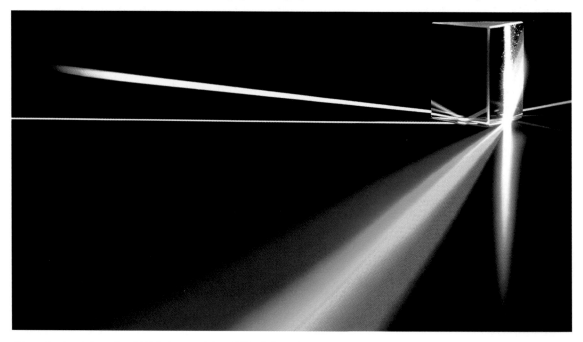

Above: A prism refracts (bends) light into a rainbow. White light, including sunlight, consists of a rainbow spectrum of colors. These colors are revealed by refraction, spreading out the light beam into the colors of the rainbow, because each color is bent at a different angle. Top left: Sunlight is scattered when it collides with airborne particles such as dust, in a process known as the Tyndall effect.

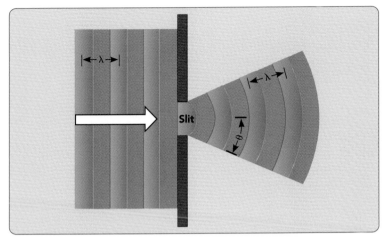

Diffraction is the process by which light passing through a narrow aperture whose diameter is close to the wavelength of light is spread out, as shown in this diagram.

flat planar surfaces whereas water droplets have curved surfaces, affecting the direction in which each will reflect or focus sunlight.

REFRACTION

Refraction, familiar to anyone who has played with a lens or prism, occurs when a ray of light enters glass, water, or ice—all of which are denser than the surrounding air. In all instances, the light ray is refracted (bent) because the speed of light in the denser medium is slower than in the surrounding medium. Refraction is what makes a drinking straw held at an oblique angle (that is, not vertically) in a glass of water appear to be bent downward. Moreover, if you could sight up the bent ray of light—just like sighting up the straw—refraction would make a star or the Sun appear higher in the sky than it really is.

White light appears to our eyes to have no color. Refraction reminds us of the opposite: that white light is actually made up of

the rainbow spectrum of colors. Shorter wavelengths of light (such as blue) are refracted, or bent, more than longer wavelengths (such as red). Thus, a prism spreads white light into a whole rainbow of colors.

SCATTERING

Scattering happens when rays of light travel through a mass of fine particles, such as dust or other aerosols. After colliding with a particle, a ray of light is altered

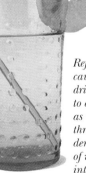

Refraction causes a drinking straw to appear bent as it passes through the denser medium of water into air.

in direction but is not altered in wavelength, so it doesn't break into a rainbow of colors. Scattering increases with increasing size of particles, becoming noticeable when viewing the beam at a right angle, and when particles are larger than about 100 nanometers (about the wavelength of far ultraviolet radiation).

First described by the British physicist John Tyndall in 1868, and mathematically analyzed by Lord Rayleigh (John W. Strutt) in 1871, such scattering is known as the Tyndall effect or Rayleigh scattering. The Tyndall effect accounts for sunbeams visible in a forest clearing or in a darkened room where dust motes are floating in the air. Rayleigh scattering also accounts for why the sky is blue and curling cigarette smoke appears bluish: shorter blue wavelengths of light are scattered some 16 times more than red wavelengths.

DIFFRACTION

Whenever a beam of light grazes the edge of an obstacle or is filtered through a narrow aperture, the beam is diffracted, that is, spread out. Optical diffraction, which is simply interference resulting from the very wave nature of light, explains why objects illuminated by point sources do not cast sharp shadows. In meteorological optics, diffraction accounts for the colors we see when water droplets, ice crystals, dust motes, or other particulates are so tiny their diameters are virtually equal to a wavelength of light—far too tiny for refraction.

The Illusionist Atmosphere

Even in the absence of any clouds or dust, pure air itself—which has thickness and density—acts like a lens, a prism, or even a mirror, through the process of atmospheric refraction.

MIRAGES

Close to the ground over sun-beaten desert sands or ice-covered oceans or lakes, the normal gradation in atmospheric density is often altered by local variations in air temperature. Such local variations in air density bend rays of light, just as light is bent when passing from air into the dense glass of a prism and then back out again. The resulting refraction creates mirages—unusual images of distant objects.

One of the most familiar mirages is the ever-receding distant pool of water that seems to appear on a hot, sun-baked highway or parched desert sands. That is an example of a so-called inferior (lower) mirage, in which an image of the blue sky overhead is seen lower than the actual sky. Inferior mirages occur when air temperature decreases with height, causing air density to increase with height. Light from the sky coming down through the air is bent upward again to an observer's eyes, and

the bright patch of reflected light looks like water. When the temperature and density gradient is especially sharp, the image is inverted, lending even more realism to an inferior mirage, such as the upside-down reflection of a car in the apparent puddle on hot pavement.

Over ice floes or bodies of relatively cold water, one may see a second type of mirage: the so-called superior (upper) or "looming" mirage, in which the image of a distant object is seen displaced upward from the actual object. A distant ship or mountain may appear to be raised up in the sky, even

if the real ship or mountain is actually below the horizon. Superior mirages occur when the temperature of the air increases with height, and thus its density decreases with height: Light that would otherwise pass overhead is bent down to the eye. Sometimes the image is magnified, displaced upward, and erect, although when the temperature gradient is sharp, the image is inverted.

When the conditions for both inferior and superior mirages occur close together, it is possible to see a rare type of combination mirage called a fata morgana (named for Morgan

Above: This pool of water and the reflection of the car are actually inferior mirages, which often occur on hot roads or desert sands. This type of mirage is created by light refracting (bending upward) as it encounters less dense air close to the ground. Top left: A Fata Morgana in the Sahara desert. The mountain range appears to float in the air, but is actually below the horizon.

Left: An engraving from French artist Camille Flammarion's The Atmosphere, *published in 1888, shows a famous mirage that appeared in Paris in December 1869. The image of the bridge is superimposed and inverted above the actual bridge, due to a sharp temperature gradient. Bottom: A rarely seen green flash, photographed in rapid succession at the last moment of a sunset.*

le Fay, King Arthur's half-sister and a shape-shifting sorceress). In a fata morgana an inferior mirage (in which the false image appears below the true position) combines with a superior mirage (in which the false image appears above the true position) to produce a double image with one image inverted below the other. The effect vertically elongates any distant mountains, coastline, or buildings to make them look startlingly like castles suspended in the sky.

GREEN FLASH

Pure air behaves like a prism as well as a mirror or lens, dispersing white light into a rainbow of colors. In the case of atmospheric refraction, the colors are dispersed vertically, because air density varies with height. Thus, near the horizon the apparent image of the setting or rising Sun is really a set of overlapping images at different colors and locations, with the redder images

lower and the bluer images higher. Usually the blue images are not visible because Rayleigh scattering from airborne dust and aerosols eliminates blue light, while water vapor absorbs the yellow light. The remaining colors are red (common in many sunsets) and green (rarely seen).

But every now and then, in perfectly clear air over an unobstructed horizon (ideally water), one can see an emerald-colored gleam of light appear at the last instant the final moment the Sun sets in the evening or the first moment the Sun rises in the morning. This green flash is caused by

a combination of the prismatic effect of atmospheric refraction, combined with mirage effects that intensify its brightness and magnify its vertical extent.

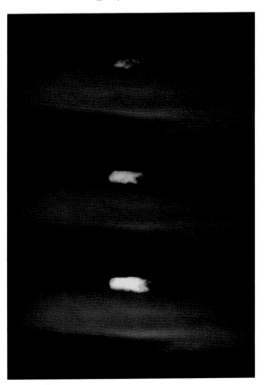

Lovely Is the Evening

There's more to air than meets the eye. In addition to the invisible molecules of different gases, air is also composed of many other substances which float suspended, largely unnoticed. These include microscopic droplets of water or ice crystals of different sizes (making up clouds or fog), motes of dust, smoke, or ash from forest fires or volcanic eruptions, grains of pollen, particulates of smog, and other materials, all of which can scatter sunlight.

SUNSET COLORS

Ever wonder why sunsets are orange and red? When the Sun is near the horizon, sunlight travels through a greater thickness of air before reaching the eye. Rayleigh scattering removes enough light from the blue end of the spectrum to make the setting Sun look orange or red. Sunsets are redder than sunrises, because by day's end humans have increased the dustiness of the air, and windy afternoons have mixed the atmosphere. If the atmosphere also holds smog, smoke particles from massive forest fires, or airborne ash from volcanoes, sunsets are particularly red, and the Sun may even appear red during the middle of the day.

CREPUSCULAR RAYS

Crepuscular rays may be familiar from artistic representations, but the bright sunbeams or bluish shafts of shadow cast by the Sun behind clouds are readily visible in real life as well. Usually appearing near the western horizon around sunset or the eastern horizon around sunrise, these painterly effects are known as crepuscular rays, the word crepuscular meaning "of or like twilight."

Although most commonly a twilight phenomenon, crepuscular rays are also readily seen midday when the Sun is behind higher clouds; in this case,

Above: Crepuscular rays, whose name comes from the Latinate name for the twilight, are a result of light scattering. Often seen around the sunset, these rays also appear midday, as shown above, when the Sun is behind high clouds. Top left: An urban skyline is silhouetted against a dramatic sunset.

the rays seem to radiate in all directions, including downward from the cloud to the horizon. Sometimes crepuscular rays can be seen crossing the entire sky and converging at the opposite horizon, a display called anti-crepuscular rays. Alternatively, after sunset or before sunrise an anticrepuscular ray may brightly illuminate a solitary patch of cloud on the opposite horizon from the Sun in an otherwise rather dark twilit sky. The apparent convergence of crepuscular or anticrepuscular rays is due to perspective, just as parallel railroad tracks appear to converge in the distance.

Crepuscular rays are more common in puffier lower-altitude cumulus clouds, although they can also appear in altocumulus. They may be seen any time of year. Watch for especially dramatic displays in the weeks or months following the

eruption of a major volcano anywhere in the world or of major long-lasting wildfires in western states, which particulates that scatter light in the stratosphere and upper troposphere.

EARTH'S SHADOW

Some cloudless dawn, face west as the Sun rises behind you. Against the invisible motes of aerosols suspended in the apparently clear air, the hemispherical shadow of the Earth may be projected as a wedge-shaped band of darker sky above the horizon. As the minutes pass, the band will lower until distant hills or buildings are illuminated by the Sun finally peering above the horizon behind you.

A sun halo appears to emanate from a woman's hand. A wide lens allowed the photographer to capture the full range of the phenomenon.

PHOTOGRAPHING ATMOSPHERIC DISPLAYS

Photographing daytime meteorological optics is far more challenging than merely spotting them. The displays can be transient, sometimes disappearing in a matter of seconds. Best advice: carry a decent camera at all times. Know the camera well; some digital cameras and most point-and-shoot cameras inevitably overexpose the phenomena, not capturing the true hues and subtle structure in full glory, and they rarely have a field of view wide enough to capture the whole display.

For most halos around the Sun, stand in a position where a street lamp or a person's hand can block the blinding image of the Sun itself, taking care to avoid lens flare (hexagonal artifacts on the film from the lens iris). For aesthetics, try to avoid letting electrical wires cross the field of view (harder than it sounds at a moment's notice). Make sure the camera is focused on infinity. Shoot as many photos as you can, widely bracketing the exposures, as underexposing is often the key to the best pictures. And stick around until the phenomenon are clearly over. Often rainbows or ice crystal displays dramatically brighten as the Sun unexpectedly reappears from behind clouds, and the shape of a sun pillar dramatically evolves in the few minutes of sunrise or sunset.

Earth's shadow is visible in the foreground of this image, photographed at dawn. The photographer was facing west as the Sun rose.

Water Droplets: Refraction and Reflection

Individual droplets of water, if relatively large (the size of raindrops), can reflect sunlight like a mirror as well as refract and disperse it into colors like a prism. In large droplets the reflection usually occurs at the inside rear surface (called total internal reflection, it's the same phenomenon that allows a prism to be used as a mirror in any pair of binoculars). Thus, water droplets can both reflect and refract light.

RAINBOW

Rainbows can be seen in the sky after a passing thunderstorm, in mist or fog, or in the spray from a waterfall or fountain. But don't wait for nature: you can make do-it-yourself rainbows from an ordinary garden hose. The most vivid colors are produced by relatively large, uniformly sized droplets illuminated by full sunlight. Refraction disperses sunlight into a continuum of all the colors of the spectrum: red, orange, yellow, green, blue, indigo, and violet (schoolchildren remember the sequence of colors by the name of an imaginary friend: Roy G. Biv). Instruments have further verified that there is a weak infra-red band beyond the red and a weak ultraviolet band beyond the violet.

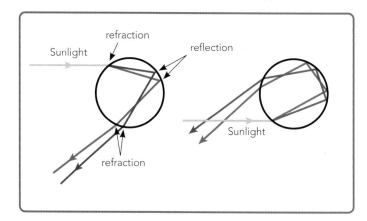

Above: A rainbow can be formed from the spray of a common garden hose. Bottom: This diagram shows how a rainbow is created. Sunlight passes into the water droplet, and is refracted, or bent; it is then reflected back out of the droplet in its constituent colors. Red and blue rays are shown here. A double rainbow occurs when there are two internal reflections: the light bounces twice. Top left: The inner bow of this double rainbow photographed in New Zealand appears almost white.

The height of the rainbow depends on the altitude of the Sun. A rainbow always appears directly opposite the Sun, with a radius of about 42 degrees centered on the antisolar point. Thus, the Sun must be lower than 42 degrees for a rainbow to be seen above the horizon. Indeed, the lower the Sun, the higher the rainbow and the more of it that can be seen, so that some summer rainbows may be seen as late as 9:00 PM. If viewed in the mist from a garden hose, a rainbow can be seen to be a complete circle.

DOUBLE RAINBOWS

Brilliant rainbows often appear as a double rainbow. The bright primary bow has red on the outside and violet on the inside, and a larger, dimmer concentric secondary bow with the colors reversed (red on the inside and violet on the outside) at an angle of roughly 51 degrees. The primary bow is produced by one internal reflection within each raindrop and the secondary by two internal reflections. Because of light lost through the double internal reflection, the secondary bow is always much fainter than the primary.

Raindrops scatter light as well as refract it, making the sky in the interior of the primary bow brighter than the sky outside of it. Similarly, scattered light makes the

The additional bands of light that appear inside this very bright rainbow are caused by light being diffracted through extremely fine water droplets. These additional bands are called supernumerary rainbows.

sky outside of the secondary bow brighter than the sky between the two bows.

SUPERNUMERARY RAINBOWS

In garden-hose rainbows or in exceptionally bright large rainbows, it is possible to see several narrow, faint, colorless

bands inside the primary bow and even outside the secondary bow. Called supernumerary rainbows or interference rainbows, they are a diffraction pattern of sunlight that results from the wave nature of light passing through extremely fine water droplets (see discussion of diffraction on the following page).

A yellow rainbow photographed in Namibia.

MONOCHROMATIC RAINBOWS

Not all rainbows show all the colors of the rainbow. Very close to sunset, when the rays of the setting Sun are distinctly red, a rainbow will be pure red instead of multicolored. Rainbows seen in fog may appear whitish because fog droplets are so tiny that light wave interference (diffraction) increases, causing wavelengths of light to overlap and produce rainbows that are milky white. Rainbows created by moonlight have colors that are so pale that the eye may perceive them as shades of gray.

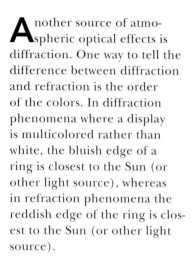

Water Droplets: Diffraction

Above: A Brocken specter, seen as concentric rings of rainbow-colored light, was photographed in Grand Canyon National Park, Arizona. An observer must be above the clouds to see this phenomenon, such as on a mountaintop or in an airplane. Top left: A partial lunar eclipse, showing a prominent corona around the Moon.

Another source of atmospheric optical effects is diffraction. One way to tell the difference between diffraction and refraction is the order of the colors. In diffraction phenomena where a display is multicolored rather than white, the bluish edge of a ring is closest to the Sun (or other light source), whereas in refraction phenomena the reddish edge of the ring is closest to the Sun (or other light source).

THE GLORY

The glory is a series of brilliant rainbow-colored rings, which are occasionally seen by airline passengers. The rings surround the shadow cast by the aircraft on a cloud or fog bank below, when the Sun is above and at the observer's back.

Glories arise when sunlight is diffracted by the fine water droplets in the clouds. The size of the rings depends both on the wavelength of light and on the size of the water droplets, with smaller droplets producing larger rings. It is most common to see just one ring, but several concentric rings can be seen when the droplets are of uniform size; up to five rings have been documented in photographs.

SPECTER OF THE BROCKEN

Related to the glory, the "specter of the Brocken" is named after the tallest peak in the Harz Mountains in central Germany, where the phenomenon is frequently seen. A person high on a mountain peak may see their shadow cast on a cloud or fog bank below when the Sun is relatively low in the sky and at the observer's back. Sometimes, the shadow looks enormous with the tiny head surrounded by a series of brilliant rainbow-colored rings, called the Brocken bow.

The shadow's huge size is basically an optical illusion. The rings, however, are caused by the same mechanism as the glory: sunlight being diffracted by fine water droplets in the clouds.

METEOROLOGICAL CORONA

The meteorological corona shares only its name with the corona that is the Sun's outer atmosphere. Otherwise, the two have nothing in common. A meteorological corona is a sequence of concentric colored rings observed around the Sun or Moon when those bodies are viewed through thin clouds—most commonly altocumulus—composed of water droplets, ice crystals, or other particles such as pollen.

Coronal rings are easiest to see around a full Moon. They also appear around streetlights in a fog, or lights viewed through moisture on misted window panes. The coronal rings are usually less than 10 degrees in diameter, about 20 times the angular diameter of the full Moon in the sky. (Astronomers consider the entire sky from horizon to zenith to opposite horizon to have an angle of 180 degrees, or half the 360 degrees of the celestial sphere.) In coronal rings, there is most often just a bright bluish white area in the center (the aureole) surrounded by a pale brownish ring. Sometimes, however, there may be a well-defined sequence of color that goes from white in the corona's center outward through blue, green, yellow, and red. Rarely, the sequence may repeat again in larger, fainter rings.

Iridescent clouds appear close to the Sun in high cloud layers, such as these altocumulus.

IRIDESCENT CLOUDS

Sometimes a common cloud appears to glow with almost metallic-looking pastel pink, purple, and green, akin to the iridescent colors seen on a soap bubble. Indeed, as with a soap bubble, cloud iridescence is an interference phenomenon, arising when sunlight is diffracted by passing through exceptionally small cloud droplets, such as those evaporating around the edges of clouds. The brightest iridescent clouds usually appear within only a few degrees from the Sun. These clouds are so dazzling that they are best viewed by looking down into the reflection from a still puddle of water, which reduces the Sun's glare.

The specter of the Brocken is essentially a glory that appears around the greatly magnified shadow of the observer. The shadow can appear to move spontaneously, as cloud layers shift.

Ice-Crystal Displays

Above: Sun dogs, or parhelia, appear on either side of the Sun, which is beginning to form a sun pillar. This phenomenon is caused by refraction and reflection. Top left: Very bright sun dogs, photographed at sunset in New Mexico. Rainbow colors can form in sun dogs, depending on the orientation of the ice crystals that form them.

A host of marvelous atmospheric optics displays arise when sunlight (or even moonlight) is refracted, reflected, and diffracted from millions of tiny jewel-like ice crystals high in the atmosphere. Ice-crystal displays are most common in cirrus or altostratus clouds. The ice crystals are almost always flat hexagonal plates (plate crystals) or long hexagonal columns (pencil crystals).

The specific display seen depends on the altitude of the Sun or Moon, the altitude affecting the angles at which light can enter and leave the ice crystals. The display also depends on the size and shape of the crystals, their optical perfection, and whether the crystals are tumbling randomly (producing white halos) or falling through the air in a preferred orientation (producing rainbow colors).

HALOS

Although in common usage the word halo usually means a circle, in meteorology a halo is a white or colorful ring, arc, streak, or spot of any shape seen in cirrus or other thin clouds or ice fogs. Although most common and most spectacular in winter, halos can be seen any time because the Earth's upper troposphere is cold enough to freeze water year-round.

The most common halo is a white or slightly rainbow-colored ring centered on the Sun or Moon, and separated from it by an angle of 22 degrees. Far less common is a larger, 46-degree ring around the Sun, an arc tangent to the upper edge of the 22-degree halo or other complex arcs in the sky.

SUN DOGS (PARHELIA)

Often called sun dogs or mock suns because their brilliance can sometimes rival that of the Sun itself, parhelia (singular, parhelion) most commonly appear as bright spots flanking the Sun at the same altitude. They may be white, but often they are rainbow-colored with the red edge closest to the Sun. When formed by moonlight, they are known as paraselenae (singular, paraselene); most reports of paraselenae have come from explorers who spot them during the long Arctic night.

Like other halos, parhelia are caused by the refraction and reflection of sunlight by tiny hexagonal ice crystals in high-altitude cirrus clouds. They can appear when the Sun is at an altitude lower than 61 degrees, but are most prominent when the Sun is low. At low Sun altitudes the parhelia appear to touch or merge with the 22-degree halo (if it also is visible); at higher Sun altitudes the parhelia grow fainter and farther away from the Sun. When parhelia in the afternoon are stretched vertically, it may signal that sunset might be graced by a sun pillar.

SUN PILLARS

A sun pillar, seen at sunrise or sunset, is a scarlet or gold shaft of light extending vertically above the Sun. The color of the pillar is the same as the Sun, and can extend from a mere 5 degrees to a spectacular 20 degrees or so above the horizon, especially when the Sun is only about 1 or 2 degrees above the horizon. Pillars have also been seen above the Moon and the planet Venus. In regions cold enough for snow and ice fogs, pillars can also be seen to form above artificial street lamps; moreover, from the air, pillars can also be seen extending below sources of light.

Pillars usually form when light from the low-altitude Sun (or other light source) reflects off the bottom horizontal surfaces of plate-shaped hexagonal ice crystals floating horizontally in the air.

Left: A sun halo encircles the subject of this photograph of unusually spectacular ice-crystal displays. Also visible are sun dogs at the left of the frame, the parhelic circle, seen as a horizontal line near the subject's shoulder, and an upper tangent arc at the top of the frame. The South Pole station can be seen in the background. Right: Light pillars caused by bright work lights near the horizon.

Special Displays

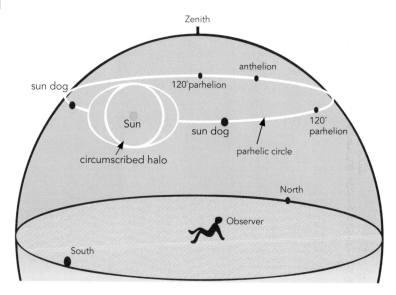

Although parhelia and other halos can be seen as often as several times a week in colder months, some ice-crystal displays have a reputation for being rare. Nonetheless, a vigilant observer of the daytime sky—especially in New England, the Mid-Atlantic, or upper Midwestern states—can look forward to seeing one of these supposedly rare ice-crystal displays once a year or even as often as once a month, more frequently in colder months. To maximize your chances of spotting such a display, make a point of stepping outside often on bright sunny days when blue skies are graced with wispy high cirrus.

Some of the ice-crystal displays—especially the parhelic circle and the circumhorizontal arc—occupy such a vast area of sky that they are impossible to photograph in full with common wide-angle lenses.

PARHELIC CIRCLE

Sometimes whitish parhelia appear with bright bluish-white tails extending horizontally away from the Sun. Those tails are actually sections of the parhelic circle, a white, horizontal circular halo that is sometimes seen to extend all the way around the sky parallel

Above: This diagram shows the placement in the sky of various atmospheric phenomena in relation to an observer facing southwest. Bottom: A circumscribed 22 degree halo at left of frame, and parhelic circle at right. Top left: The lower portion of a circumzenithal arc appears to form a smile in the sky.

to the horizon at exactly the same altitude as the Sun.

Other sun dogs can appear along the parhelic circle. One of these is the anthelion, a pale white spot a couple of degrees across and directly opposite the Sun. Another pair of sun dogs are the paranthelia, which occur on the parhelic circle 120 degrees away around the sky. They are pale, colorless (white) spots perhaps a degree across, marking locations where other fainter arcs of halos cross the parhelic circle.

CIRCUMZENITHAL ARC

The circumzenithal arc is a rainbow-colored arc centered on the zenith with its bow directly above the Sun and its reddish band toward the Sun. Although sometimes stunning in its colorful brilliance, this arc often escapes people's notice because it is vertically overhead. It can be seen any time of the year and is often accompanied by parhelia and the 22-degree halo, although sometimes it also appears alone. Because of the geometry of the rays that produce it, the circumzenithal arc can appear only when the Sun is lower than an altitude of 32 degrees above the horizon; it is narrow and well defined when the Sun altitude is 22 degrees, appearing more diffuse at other Sun altitudes.

CIRCUMHORIZONTAL ARC

A complementary ice-crystal display is the circumhorizontal arc, which can appear only when the Sun altitude is higher than 58 degrees (making it viewable only near the summer solstice in the northern United States, southern Europe, and Japan). Sometimes with stunning colors, it is a truly huge display. Looking like a fat, diffuse rainbow absolutely parallel to the horizon, the circumhorizontal arc can extend almost a third of the way around the horizon, centered under the Sun with its reddish band above (toward the Sun).

Like the circumzenithal arc, it is often seen at the same time as the 22-degree halo. It cannot, however, be seen at the same time as the circumzenithal arc.

This spectacular green aurora is the result of charged particles from the solar wind colliding with Earth's magnetosphere. Auroras are most commonly visible from latitudes close to the poles, though the most spectacular auroras are sometimes seen in the middle latitudes and even near the Equator as well.

VISIBLE ATMOSPHERIC CHEMISTRY

The nighttime atmosphere also has its glories, chief among them being the aurora—the dancing northern or southern lights. Aurorae are produced when charged particles from the Sun encounter Earth's magnetic field and spiral down toward the poles, colliding with air molecules in the ionosphere. The air molecules glow from the collisions like the gas in a neon sign, flickering in lurid greens (usually produced by oxygen) and reds (usually produced by nitrogen). Auroras are most numerous and spectacular when the Sun's face is peppered with great numbers of sunspots. Although most commonly seen from high north or south latitudes, occasionally an auroral display can be seen from the tropics or the Equator.

Also, recall that beautiful meteors—commonly called falling or shooting stars—are an upper-atmosphere phenomenon. Meteors are visible evidence of the air protecting us on the ground from grains of cosmic sand and gravel speeding through interplanetary space at tens of miles per second. The meteorite collides with molecules, and is heated by atmospheric drag until it vaporizes, leaving behind a train of glowing air molecules.

CHAPTER 11

WEATHER ON OTHER PLANETS

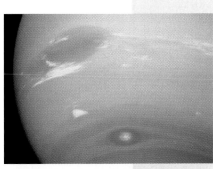

Left: NASA's Voyager 1 *took this photograph of Jupiter's Great Red Spot, an anti-cyclonic storm more than three times the size of the Earth. The white oval below it is a smaller storm composed of cool clouds higher in the atmosphere. Top: Hurricanes on Neptune. At the equator is the Great Dark Spot with the white area that* Voyager 2 *scientists nicknamed "Scooter" directly below it. Further south and to the right is the Dark Spot 2, which has a bright core. Bottom: Saturn's C-ring in purple and B-ring in yellow, photographed through ultraviolet, clear, and green filters by NASA's* Voyager 2. *These rings are composed of hundreds of thousands of ice crystals.*

All the planets in our solar system—indeed, all matter in the universe—is made up of the same chemical elements as Earth. The other planets in the solar system differ so from Earth and from one another principally because of their different masses and proximity to the Sun.

Earth is not the only planet with weather. Most of the other planets and even some of the planetary satellites have atmospheres. And some also have far more impressive weather than Earth.

Although both Mercury and the Moon have an extremely low surface pressure, both still have a distinct exosphere. Moreover, the Moon may also have an "atmosphere" of restlessly moving motes of electrostatically charged dust.

Hothouse Venus is blanketed with a hot, crushing, acidic atmosphere as dense as a deep ocean. Mars's carbon-dioxide atmosphere, on the other hand, is thin and arid, punctuated by planet-wide summer dust storms that can last for months.

The gas giant Jupiter has had a hurricane raging for some three centuries. Tranquil-looking pastel Saturn has lightning and auroras, and its very rings are surrounded with their own atmosphere of oxygen.

The apparent twins Uranus and Neptune have their own icy seasons, and the frigid, dense atmosphere of Pluto, the dwarf planet, appears to be expanding instead.

Airless Extremes of the Moon and Mercury

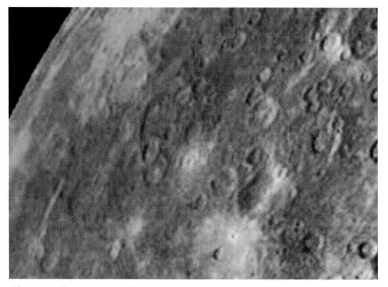

Above: A color composite of Mercury's surface. Scientists believe that the elements in Mercury's atmosphere are supplemented when meteorites, energetic subatomic particles, and photons from the Sun hit Mercury's surface and vaporize rocks. Top left: This image of the Moon from the Galileo *spacecraft shows dark areas that are impact basins filled with lava rock. The rayed lighter area at the bottom of the Moon is the Tycho impact basin. Craters at the lunar poles may harbor frozen water or ices of volatile compounds.*

Although both the Moon and Mercury are considered to be airless, they still have a tenuous outer exosphere with a surprising variety of elements. Moreover, the Moon may be surrounded with a restless "atmosphere" of leaping dust particles.

MERCURY

The only time human eyes have seen Mercury from up close was through the cameras of the *Mariner 10* spacecraft in 1974–75. So much is still unknown about the planet closest to the Sun. Superficially, cratered Mercury looks much like the Moon, which it also resembles in size. But Mercury's much-stronger gravitational field indicates that the planet is almost as dense as Earth, suggesting it is composed of mostly iron instead of mostly rock.

Mercury's tenuous exosphere—so thin that its atoms and molecules can never collide—extends some 600 miles (about 1,000 km) above the surface. In it, measurements from both *Mariner 10* and half a dozen ground-based telescopes have found traces of helium (likely from the solar wind), sodium, argon, neon, hydrogen, potassium, and even oxygen—in fact, oxygen may account for more than 40 percent of Mercury's exosphere. Although there are many uncertainties,

scientists believe that most of the elements in Mercury's exosphere originate when meteorites, energetic subatomic particles, and photons (packets of light) from the Sun strike the planet's surface, vaporizing rocks.

Because Mercury is only 36 million miles (58 million km) from the Sun, the daytime temperature on its sunlit side soars above 840°F (450°C)—hot enough to melt lead. Still, echoes from Earth-based radar suggest that ices of different frozen

volatiles—water among them—might exist in its polar regions in permanently shaded craters.

THE MOON

When early astronomers realized that the Moon was a separate planetary body, the tantalizing question arose: did the Moon also have an atmosphere? If the Moon did have an atmosphere, might it be populated by living beings? These questions remained unanswered from the sixteenth well into the twentieth century.

Finally, instruments carried to the Moon by the astronauts of *Apollo* missions *11, 12, 14, 15, 16,* and *17* between 1969 and 1972 were able to make definitive measurements, supplemented by later spacecraft and ground-based observations. As far as we know, the answer is yes for the atmosphere—counting its exosphere as a proper atmosphere—and no for living beings.

The Moon's exosphere consists of nearly equal proportions of hydrogen, helium, and neon, laced with dashes of sodium and potassium and some heavier molecules: carbon dioxide, methane, ammonia, and water vapor. Because lunar gravity is so low— only a sixth that of Earth—solar heat can accelerate most atoms and molecules to speeds that allow them to escape, unlike on Earth, which is massive enough that its gravity can hold onto most of our atmosphere. Thus, the molecules in the Moon's exosphere have to be replenished from somewhere. The solar wind bathes the lunar surface with many types of charged subatomic particles; impacts from meteorites and comets also make their contributions. Seismic activity may also release helium and argon from deep inside the Moon.

The Moon's exosphere is very dusty, revealed in part by several Apollo astronauts who sighted "twilight rays" that resemble crepuscular rays on Earth (see chapter 10). Other measurements by various instruments have confirmed the existence of suspended, moving dust particles above the Moon's surface. Some astronomers now hypothesize that both the solar wind and energetic ultraviolet radiation from the Sun may ionize (strip outer electrons from) the top layers of the

Astronauts orbiting the Moon on Apollo 17 *observed and sketched these rays that occurred during lunar sunset and sunrises. Variously called "bands," "streamers," and "twilight rays," by the astronauts, they appear to be similar to Earth's crepuscular rays and may be caused by sunlight being scattered by electrostatically charged dust above the Moon.*

lunar regolith, the Moon's fine powdery soil. The regolith builds up a static charge until fine dust particles are repelled, being launched tens or even hundreds of kilometers above the surface. Unlike the gaseous molecules, the dust particles do not reach escape velocity, so eventually they fall back to the surface—but are repelled again and again. Thus, the Moon may be surrounded by a kind of atmosphere of moving dust particles.

Like Mercury, the Moon has deep craters in its polar regions that never see the light of Sun.

The Discovery Rupes, the fault line in the center of the above image, is thought to be formed as Mercury cooled and grew smaller.

Hothouse Venus

In many ways, Venus seems to be Earth's twin; they are nearly the same size, for starters. The two planets' temperatures tell a different story, though; Venus is truly an inferno compared to the life-giving haven of Earth. Instruments on every spacecraft that has landed on Venus have stopped working after close to an hour. The atmosphere is so hot, dense, and caustic that despite every protection humans can design, the spacecraft have been baked, crushed, and dissolved.

POISONOUS GREENHOUSE

At 7,500 miles (12,100 km) across, Venus is 95 percent the diameter of Earth. Thus, until the early twentieth century some astronomers thought its blanket of air might be similar. But nothing could be farther from the truth.

Every spacecraft that has landed on Venus has found itself parachuting down to the surface through a yellowish, soupy atmosphere that is more than 95 percent carbon dioxide and less than 5 percent nitrogen and trace elements and compounds. They have drifted downward through clouds made of droplets of nearly pure battery acid (sulfuric acid), one of the most caustic and toxic acids known.

Once on the surface, all the spacecraft quickly succumbed to heat great enough to melt aluminum and other materials used for electronics. Venus, despite being twice as far from the Sun as Mercury, is the hottest planet in the solar system both day and night, averaging nearly 900°F (480°C). The reason is a runaway greenhouse effect; the Venusian atmosphere is composed mostly of the greenhouse gas carbon dioxide, so longer-wavelength thermal infrared radiation is effectively trapped within a highly efficient thermal blanket.

Moreover, the spacecraft have found themselves withstanding atmospheric pressure 90 to 100 times greater than Earth's air pressure at sea level—literally the same as the crushing force found some 0.6 mile (1 km) deep in Earth's oceans. How can a planet slightly smaller than Earth have such high atmospheric pressure? The answer is carbon; most of the carbon on Earth is locked into carbonate rocks, but on Venus it is free in the atmosphere. If all Earth's carbon were free in its atmosphere, its air pressure would be just as great.

BIZARRE CONTRASTS

From Earth, Venus is gorgeous: It appears as the brilliant cream-white "evening star" that often glows after the Sun has set, and as the "morning star" that precedes

Below: Radar data and altimetry as well as coloring from previous Soviet Venera spacecraft photographs of Venus's surface were used to generate a three dimensional image of the Maat Mons volcano. Volcanoes have and may continue to contribute minerals and gases to Venus's atmosphere. Top left: Ultraviolet light shows the sulfuric acid clouds of Venus's surface.

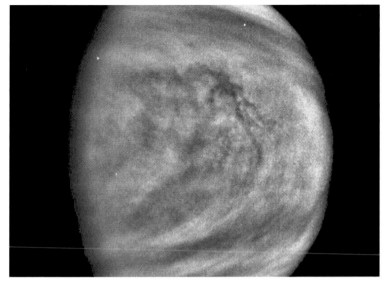

A high-pass spatial filter helps highlight contrasts in cloud features and shows the considerable convective activity in the Venus's sulfuric acid clouds.

some looking like hurricanes with two eyes—are constantly swirling in Venus's atmosphere. They effectively carry heat from the planet's equator to its poles, nearly equalizing temperatures at all latitudes as well as on both the day and night sides, regardless of the planet's slow rotation.

MANY MYSTERIES

Although Venus today has scarcely any water vapor, it may have had substantial water in the past. The main culprit may be its lack of a protective ozone layer. On Earth, the ozone layer in the stratosphere prevents solar ultraviolet from penetrating down to the troposphere and destroying water vapor.

On Venus, however, there is no ozone layer, so ultraviolet radiation penetrates to within a few miles (kilometers) of the surface, dissociating any atmospheric water vapor into its constituent hydrogen and oxygen. In such a hot environment, the lightweight hydrogen would easily escape into space, and the remaining oxygen combined with carbon to form carbon dioxide and carbon monoxide, with a permanent loss of water to the planet.

the Sun, rising before dawn. Through a telescope or from a spacecraft camera at visible wavelengths, Venus also looks like nothing more than a pale beige sphere.

Spacecraft that landed on Venus revealed that its lower atmosphere is stifling, hardly stirred by slow movements of just 0.5 to 2 miles per hour (1 to 4 km/h), where gentle zephyrs at the surface are more like deep-sea currents. But close-up observations from orbit in ultraviolet light

reveal another story in its upper atmosphere. Viewed in ultraviolet light, the apparently featureless clouds show wide dark bands and belts that restlessly form, swirl, and dissipate as upper-level winds speed at nearly 200 miles per hour (300 km/h). In what meteorologists call "super-rotation," the high-speed winds circle the planet in just four days, more than 29 times faster than its solar day (sunrise to sunrise) of 117 days.

Cyclones some 60 to 300 miles (100 to 500 km) across—

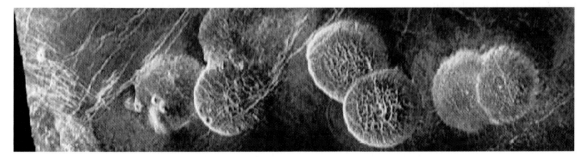

Volcanic domes formed in the Alfa Regio region of Venus. Venus's lower atmosphere is stiflingly still, perhaps stirred only occasionally by events like the slow steady eruptions that formed these "pancake" domes.

Wind-Swept Mars

Mars, Earth's other nearest neighbor and at 4,220 miles (6,790 km) across, just over half Earth's diameter, is still thought to be a possible environment where primitive forms of life could have evolved. Like Venus, its atmosphere is more than 95 percent carbon dioxide, with traces of nitrogen, argon, and even water vapor. Unlike Venus, however, Mars's atmosphere is so thin that its surface pressure is less than 1 percent that of Earth at sea level, about the same as Earth's stratosphere at some 20 miles (35 km) altitude, precluding any warming greenhouse effect. Moreover, because Mars is half again as far from the Sun as Earth, the surface is frigid and dry.

Nonetheless, the planet has fascinating weather.

Above: This infrared photo captured by NASA's Mars Odyssey *shows layering on the surface of Mars. The brightness of the image indicates surface temperatures that range from -4°F to 32°F (-20°C to 0°C). Temperature differences are due to composition of the layers as well as their orientation toward the Sun. Top left: The southern hemisphere of Mars has a seasonal polar carbon dioxide frost cap in this composite photo taken in May 2003 by the* Mars Orbiter *camera. A dust storm can also be seen as a darker area at the upper right on the Acidalia planes.*

TWO SUMMER CLIMATES

Although the axial tilt of Mars is similar to that of Earth (25.2° versus 23.5°), the orbit of Mars is very different. It is so elliptical that Mars is a full 27 million miles (43 million km) closer to the Sun at perihelion than at aphelion. Because perihelion coincides with the southern-hemisphere Martian summer, southern summer (158 days) is significantly shorter and warmer than northern summer (183 days). That leads to strikingly different global weather.

In the warmer southern summer, temperatures rise so much that water vapor can't condense into clouds. Solar heating at the surface starts convection and spinning dust devils, some of which tower as high as 5 or 6 miles (7 to 10 km)—taller than any terrestrial tornado. The dust devils throw dust high into the atmosphere, which insulates and further warms the surface. Some years, the dust becomes whipped around the entire planet in global dust storms that can persist for months. The southern ice cap, principally of dry ice (frozen carbon dioxide), sublimates directly from solid into gas, increasing the planet's atmospheric pressure by 30 percent.

In the cooler, longer northern summer, temperatures stay low, so water vapor condenses into clouds that cluster around the peaks of volcanoes and further reflect sunlight. Ice also condenses on dust particles, causing them to fall to the ground. Air is clear and cold.

WHERE DID MARTIAN WATER GO?

Gullies, layered terrain, the height of the polar ice caps, and numerous other surface features on Mars look as though they have been carved by running water or are supported by polar ice water. At today's low atmospheric pressures, liquid water cannot exist on Mars, but quickly vaporizes. Many planetary scientists think Mars may have had more water in the past.

The reason Mars lacks water may have to do with the fact that it has no magnetic field. Earth's magnetic field is very strong, slowing and deflecting most of the charged particles from the solar wind, preventing most of them from affecting Earth's atmosphere or surface. The comparatively few particles that do penetrate Earth's magnetic field interact with the upper atmosphere to create the ionosphere. On Mars, however, no global magnetic field slows the solar wind, which can hit the surface directly, except over parts

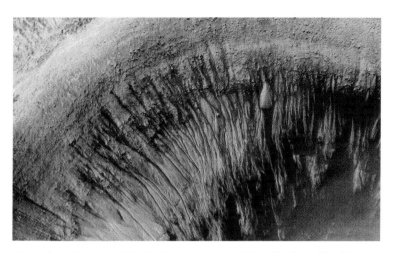

Above: A small crater within the Newton crater on Mars. The finger-like ridges are speculated to be formed from deposits carried by flows of water. Below: A panoramic view of Mars's rocky surface from the Mars pathfinder.

of Mars's southern hemisphere that have regional magnetic fields as strong as Earth's. Just over those areas, Mars has an ionosphere. Elsewhere, it is believed that the solar wind dissociates water molecules and sweeps them out into space.

OPEN QUESTION

The Martian atmosphere is strongly oxidizing (chemically reactive), acting like a strong alkali or bleach. Mars's reddish color is characteristic of various iron oxides—that is, rust, because the soil is so rich in iron. Unprotected terrestrial life would die in such a reactive environment, making some scientists wonder whether any life forms could have evolved on Mars. Moreover, recent observations suggest that electrical discharges

(not necessarily lightning per se) within some Martian dust storms might precipitate snow made of hydrogen peroxide, a chemical commonly used as a disinfectant on Earth, which could further sterilize the ground.

In contrast, recent observations have also detected trace amounts of methane in Mars's atmosphere. On Earth, methane usually has a biological origin (although it also could originate from impacts of comets), causing scientists to wonder about the viability of underground bacteria; that speculation was further fueled by the 2006 discovery of bacterial colonies in nutrient-rich groundwater nearly 2 miles (3 km) underground in South Africa, whose sole source of energy is radioactive decay from nearby rocks.

Stormy Jupiter

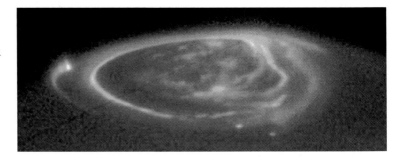

Stunning Jupiter is one of the most beautiful planets in the solar system, with its parallel dark bands and light zones moving in opposite directions around the planet, and its magnificent oval spinning Great Red Spot. Close-up photographs from the two *Voyager* spacecraft and from *Galileo* reveal stupendous detail: whorls and eddies worthy of being a work of art.

Jupiter is about 75 percent hydrogen and 25 percent helium by mass, with trace amounts of methane, ammonia, water, and sulfur compounds, which likely account for the different colors. Different colors correspond to three distinct depths of cloud layers, with reds being highest, browns and creams being lower, and blues being the lowest.

FAILED STAR?

Jupiter gives off far more heat than it receives from the Sun, because its interior is hot, some 20,000 Kelvin. In the past some scientists speculated it might be a failed star; it so, it failed by a long shot, as it would need to be about 80 times more massive to ignite thermonuclear reactions in its core. As matters stand, the heat it generates is due to the compression of gravity.

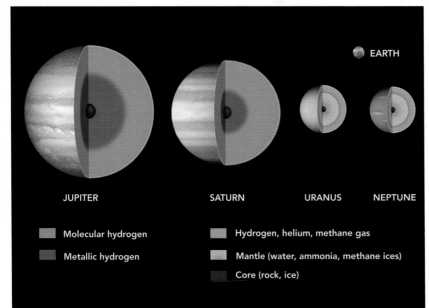

JUPITER   SATURN   URANUS   NEPTUNE

EARTH

- Molecular hydrogen
- Metallic hydrogen
- Hydrogen, helium, methane gas
- Mantle (water, ammonia, methane ices)
- Core (rock, ice)

Left: Unlike the Earth's surface, which has a hard crust, Jupiter's surface is gaseous-liquid. This makes the boundary between the planet and its atmosphere almost indistinguishable. Gravity compression, caused by Jupiter's metallic core, generates the planet's intense heat. Above: Auroras are created by charged particles from the Sun drawn by gravity toward Jupiter's poles. This phenomenon is similar to Earth's northern lights. Top left: Clouds in shades of reddish brown and light blue cover the surface of Jupiter. Jet streams run parallel to horizontal bands at speeds of up to 300 mph (480 km/h). The shadow of Jupiter's moon Europa is the dark dot at the lower left.

The *Voyager, Galileo,* and *Cassini* spacecraft observed auroras near Jupiter's polar regions that are hundreds to thousands times more intense than the northern or southern lights on Earth, even emitting pulses of X rays. The bulk of Jupiter's auroras are powered by charged particles from the solar wind, as they are on Earth. But scientists have observed bright streaks or spots in Jupiter's auroras which they believe are caused by magnetic flow between the planet and its satellites. Jupiter's moon Io in particular gives off a great deal of charged particles, apparently from its constantly erupting volcanoes.

Lightning also flickers in Jupiter's upper atmosphere, apparently correlated with low-pressure areas and the planet's global circulation.

RED AND WHITE SPOTS

The Great Red Spot in the planet's southern hemisphere rotates counter-clockwise, making it an anticyclone (high pressure region). As big as two Earths laid side by side, one complete rotation takes four to six days, with winds traveling as fast as 270 miles per hour (400 km/h). Long a feature of the planet, it was first reported by telescopic observers more than three centuries ago.

Smaller white ovals, which appear to be cold storms, occasionally get sucked into the Great Red Spot or redirected around it. Sometimes the white ovals also merge with one another and grow larger and stronger.

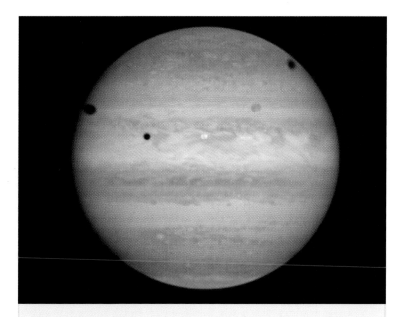

Jupiter is seen here with its satellite Io (the white circle in the center of the planet) and Ganymede (the blue circle at the upper right) and their shadows. The shadow of Callisto is visible as well. Jupiter's satellites have their own distinct atmospheres and weather.

JOVIAN SATELLITE ATMOSPHERES

Jupiter has four large satellites that can even be seen in a small telescope or pair of binoculars, as well as several dozen much smaller ones. Named the Medician stars by Galileo for his patrons, all four have atmospheres of their own.

Ganymede, Jupiter's largest satellite (and the largest satellite in the solar system), is larger than the planet Mercury. Despite that, it was long thought to have no atmosphere, although recent detections of ozone near the surface have led astronomers to discover that it has an ultrathin atmosphere of oxygen. It may also have its own auroras, possibly the first to be discovered on any planetary satellite.

Europa, Jupiter's second-largest satellite, is slightly larger than Earth's Moon. Liquid water exists under its icy surface—in fact, it may have even more water than Earth—and it is surrounded by an exceptionally tenuous oxygen atmosphere.

Callisto, the third largest satellite slightly smaller than Mercury and the most heavily cratered body in the solar system, has an exosphere of carbon dioxide.

Io, larger than the Moon, is the most seismically active body in the solar system, with hundreds of active volcanoes, some of which spew matter nearly 200 miles (300 km) above the surface. Its thin atmosphere is principally sulfur dioxide laced with potassium, oxygen, sodium, and chlorine—these last two leading scientists to wonder whether Io's violent volcanoes might be influenced by the existence of salt.

Banded Saturn

The most striking feature about Saturn is not the ball of the planet itself, but the marvelous rings surrounding it. The planetary ball, indeed, is much less colorful than Jupiter, with muted beiges and tans and indistinct features. Its atmosphere is mostly hydrogen, with some helium, ammonia, methane, and water vapor, along with sulfur compound that give the planet its yellowish tinge. But the apparent lack of sharply defined features is deceptive, a mere accident of the fact that the planet is obscured by an upper layer of thick smoggy haze.

As with Jupiter, clouds occur in distinct cloud decks at specific heights: a lowest layer of water ice at about 10 times Earth's atmospheric pressure, a middle layer of ammonium hydrosulfide at 5 times, and an upper layer of ammonia at about the same. Above those are visible cloud tops and several layers of haze.

Saturn, like Jupiter and Neptune, emits more energy than it receives from the Sun. It is only about 70 percent the density of water, meaning that if there were a bathtub large enough to hold it, it would float.

At the time the *Voyager 2* spacecraft flew by Saturn in the early 1980s, an enormous storm was raging with winds up

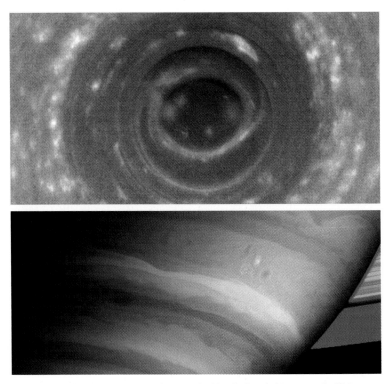

Top: A storm vortex on Saturn photographed by the Cassini *spacecraft. This storm exhibits an eye, eye-wall clouds, and spiral arms, features thought to be unique to hurricanes on Earth. The storm differs from terrestrial hurricanes in that it appears to be fixed at Saturn's south pole and is not formed over an ocean of water. Bottom: A bright, complex convective system known as the Dragon storm glimmers in Saturn's storm alley, September 2004. Top left: Saturn in 1981, with its rings.*

to 1,000 miles per hour (1,700 km/h). Such a powerful storm disturbs the planet about once every three decades, approximately once each Saturnian year.

Monumental auroral displays also crown the poles of Saturn, some extending as high as 1,200 miles (2,000 km) above the cloud tops.

Not to be outdone, the *Cassini* spacecraft in 2004 revealed that Saturn's glorious rings have their own tenuous atmosphere independent of any belonging to the planet or its satellites. Surprisingly, the rings are composed of molecular oxygen, same as in Earth's own atmosphere. Best estimations are that the oxygen is released when sunlight dissociates molecules of water vapor that sublimate from chunks of water ice that make up the rings.

SATURN'S SATELLITES

Titan, Saturn's largest satellite, is larger than Mercury and more massive than Pluto. It has an atmosphere 60 percent denser than Earth's (1.6 bar) that is more than 400 miles (700 km) deep. Although it is mostly nitrogen (80 to 90 percent), it is not breathable because its other minor constituents are argon and methane, with traces of carbon dioxide, water vapor, ethane, carbon monoxide, and a dozen other elements and compounds. There are even occasional clouds, and its rain may consist of droplets of methane or ethane; surface features such as dry riverbeds and coastlines suggest that precipitation may have fallen in the past. Titan's

An ultraviolet image of Saturn's rings suggests a distinct difference in the composition of the A ring (in red) and the B ring (in turquoise). The A ring is thought to be formed of sparse ringlets of dirt and ice and the B ring of "clean" ice in larger particles.

orange atmosphere is thick with several distinct layers of haze or photochemical smog that obscure its surface, somewhat akin to Earth's own primitive atmosphere around the time life was emerging (except for its frigid temperature).

The *Cassini* spacecraft revealed that Enceladus, Saturn's sixth-largest satellite, appears to have an atmosphere of water vapor. Because the body is so small—only 300 miles (500 km) across—its gravity could not hold onto any atmosphere; thus, scientists suspect that the atmosphere may be continuously replenished by geysers, volcanoes, or other outgassing from the moon's interior.

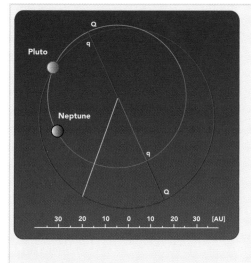

Pluto's highly elliptical orbit around the Sun takes two and a half centuries to complete.

PLUTO'S PEEKABOO ATMOSPHERE

Pluto's atmosphere is present only during part of its two-and-a-half-century-long orbit around the Sun. Even more so than Mars, the orbit of Pluto is highly elliptical—so much so that the body actually spends part of its orbit closer to the Sun than Neptune, as it did most recently between 1979 and 1999. Pluto is so distant that the Sun ranges from a summer maximum brightness of 1/900th Earth's to a winter dimness of 1/2,500th (about as bright as the full Moon on Earth). Indeed, winter temperatures are so frigid that Pluto's very air (of mostly nitrogen, like Earth's, with traces of carbon monoxide and methane) should freeze and fall to its surface like frost.

At least that is what scientists have expected as Plutonian summer has turned into autumn. Recently astronomers observing Pluto occult (eclipse) distant stars have been surprised to observe that Pluto's atmosphere seems to be expanding rather than contracting. More details will become known when the *New Horizons* spacecraft arrives at Pluto in 2015.

Uranus and Neptune: Frigid Twins

Although roughly the same size and greenish-blue color, the gas giant planets Uranus and Neptune appear to be virtual twins, even when the *Voyager 2* flew relatively close by. They both have atmospheres of hydrogen, helium, and methane, although Neptune's stratosphere also has traces of carbon monoxide. They both seem to have cores made of rock about the same size as Earth, and mantles of a kind of icy, smelly slush made of water, methane, and ammonia.

But they turn out to have significant differences in the structure of their atmospheres as well as in their weather.

FLIPPED-OUT URANUS

The most remarkable fact about Uranus is that it is essentially flipped on its side: unlike most planets, whose rotational axes are more or less perpendicular to the planes of their orbits around the Sun, Uranus's axis of rotation—tilted at 98 degrees—lies almost in its orbital plane. That means that in its northern summer its north pole points almost directly toward the Sun and vice versa for southern winter, giving rise to extreme seasons throughout its 84-year-long year (currently, its south pole is pointing almost directly at Earth).

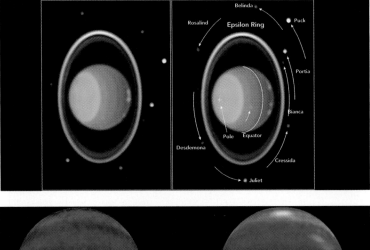

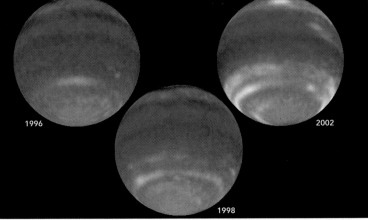

Top: The bright spots on this image of Uranus are clouds. The red coloring of the clouds on the right edge indicates that they are higher in the atmosphere while the green cloud near the equator is closer to the surface of the planet. These clouds as well as the satellites that travel around the planet show Uranus's unusual axis of rotation, almost parallel to the plane of its orbit rather than perpendicular. This gives rise to extreme seasonal variation; in the northern hemisphere's summer the pole is pointed almost directly at the Sun. Bottom: The clouds in Neptune's atmosphere, in the southern hemisphere, steadily increased in number and brightness over a six-year span as shown by the Hubble space telescopes' cameras. This might be a seasonal effect since Neptune is entering the spring of its 165 year-long orbit. Top left: This photo of Neptune taken in 1989 from the Voyager 2 *spacecraft shows the Great Dark Spot cloud system, and its bright companion, Scooter, directly below it. Winds near the Dark Spot traveled westward at up to 1,200 mph (2,000 km/h), the strongest gusts measured on any planet.*

Appearing virtually feature-less even to the powerful Hubble Space Telescope, Uranus shows few upper-level clouds. That may be changing, however; the planet entered its spring season in 2006, and a huge dark hurricane nearly the size of the United States was spotted, as well as numerous other fast-moving clouds that change shape and form other features.

DIAMOND-STUDDED NEPTUNE

Neptune is striped with distinct bands and belts parallel to its equator, just as Jupiter and Saturn are. Unlike these two planets, however, Neptune gives off about double the heat than it receives from the Sun. That leads astronomers to believe that it has some kind of internal source of heat that gives rise to convection and the equivalent of Hadley and other cells that create its banded appearance. The internal heating may even be enough

to decompose methane into its constituent carbon atom and four hydrogen atoms. According to one exotic theory backed by experiments, the carbon atoms could bind together under enough heat and pressure that solid diamonds literally might fall as hail toward its solid rocky core.

When *Voyager 2* flew by Neptune in 1989, it found a dark hurricane as large as Earth itself swirling in Neptune's atmosphere. At first scientists wondered whether it might be like Jupiter's Great Red Spot, but five years later the Neptunian storm seemed to have disappeared as it could not be seen by the Hubble Space Telescope. Nonetheless, other dark spots have occasionally appeared. Voyager also photographed long thin clouds that looked remarkably like cirrus in its upper atmosphere, and clocked winds moving as fast as 1,200 miles per hour (2,000 km/h)—the fastest in the solar system.

This streak of clouds in Neptune's atmosphere ranges from 31 to 124 miles (50 to 200 km) wide. The shadows to the right of the clouds indicate a vertical height of about 31 miles (50 km). To an observer on Neptune they would appear fluffy, like earthly cumulus clouds.

Since 1980, Neptune has clearly been getting slowly brighter, perhaps some seasonal response (with a 165-year-long orbit, a single season can last more than four decades).

NEPTUNE'S SATELLITE TRITON

Neptune's largest moon is Triton, the largest satellite to orbit a planet in a retrograde direction (opposite to the direction of the planet's rotation).

Although at 1,700 miles (2,700 km) across, Triton is only three-quarters the diameter of Earth's airless Moon, it nonetheless has a perceptible atmosphere of nitrogen and methane. *Voyager 2* also spotted distinct clouds that might be of particles of nitrogen ice crystals. *Voyager 2* also spotted dark plumes, suggesting that Triton's surface (taken to be very young in astronomical terms as it has no craters) is peppered with erupting geysers—making Triton the only body outside of Earth, Venus, and Io known to be volcanically active at the current time.

Triton also has ice caps that, since the 1989 flyby of *Voyager 2*, seem to be melting, increasing the density of its atmosphere to almost as thick as that of Mars, and causing some scientists to observe that the satellite might be experiencing its own version of global warming.

Neptune's satellite, Triton, has the coldest surface temperature of any location in the whole solar system, -391°F (38°K). Its south pole is covered by a massive ice cap, which appears pink.

CHAPTER 12

WEATHER, CLIMATE, AND SOCIETY

Left: The lights of cars glimmer in New York City's otherwise dark skyline during the August 14 blackout of 2003.
Top: Architecture represents human efforts to provide total protection from the elements. This bridge in Moscow ensures pedestrians a crossing free from exposure to rain, sleet, or snow. Bottom: Women and children wait in line to receive assistance after flooding in August 2006 displaced more than 650,000 people in Southern and Western India.

With our hermetically sealed "climate-controlled" skyscrapers and office buildings, humans like to think we wield mastery over everything, even the weather. But one summer lightning strike to an electric power plant quickly reminds us that we are not at all in charge of weather. Tornadoes, which can level houses in minutes, are similar reminders; on a larger scale, so are hurricanes, floods, and blizzards.

Regardless of its vagaries, weather is big business. Weather disasters have wiped out entire companies and individuals' life savings—but recovery from them has also brought boom times to construction contractors, architects, and lumberyards.

Weather has turned the tide of battles that helped determine the outcome of whole wars, including England's repulsion of the Spanish Armada, George Washington's winters during the American Revolution, and the D-Day invasion of World War II.

Humans may not control weather, but we certainly influence both weather and global climate by releasing pollutants into the air. Indeed, one of the biggest questions facing humanity today is: How much of the current climate change is a result of human actions? If so, what can—and should—we do about it?

Weather Is Big Business

Good or bad, weather affects the economies of states and nations both directly and indirectly. According to one estimate, as much as a third of the United States's gross domestic product—to the tune of some $3 trillion annually—is sensitive to vagaries of weather.

Weather events affect the availability, and thus the pricing, of goods and services within regions, nations, and even internationally. The chain also works in reverse: consumption habits in industrialized nations are also affecting the weather in nations half a world away.

DIRECT EFFECTS OF WEATHER

What would the economies of Colorado and Utah, or of Switzerland, be like without their world-class skiing? And balancing the thousands of tourists each year who think snow, there are thousands of others who flee northern winters in pursuit of year-round moderate temperatures, settling in second homes in Florida, the south of Spain, Cancún, or other subtropical destinations. Regardless of direction of travel, winter drives people to be customers of airlines, travel agents, resorts, restaurants, and attractions.

Top: A man holds freshly gathered coffee berries in Kenya, Africa. A poor growing season in Kenya could affect the price and quality of coffee for people a continent away. Bottom: A store display of bathing suits reflects the power of the seasons to fuel marketing trends. Top left: Skiers survey prime conditions on slopes in Telluride, Colorado. This area's economy relies heavily on winter vacationers.

Even local non-leisure businesses have seasonal peaks and valleys in their income or variations in the nature of their work as a result of weather. Landscapers primarily do plantings in spring and fall, mowing and trimming in summer, and snow removal in winter. Roofers, pavers, and exterior painters depend on relatively warm days without precipitation. And in the U.S. Midwest where repeated cycles of freezing and thawing annually degrade asphalt and open new potholes, a standing joke for highway construction crews and travelers alike is that there are only two seasons: winter, and orange-barrel season.

The pricing of many commodities varies with the weather. On commodities markets, grain and livestock futures are bought and sold at prices that are affected by expectations for the current agricultural growing season. A record-cold winter may bring price spikes in natural gas, heating oil, and fiberglass insulation—not to mention pipes for plumbing that may have frozen and burst. An unusually hot summer spikes demand for window-unit air conditioners, whose peak use can cause "brownouts" as electric utilities try to equalize the load over all their customers by cutting back a bit on voltage.

WEATHER AND THE SUPPLY CHAIN

Some of the economic influence of weather is indirect, but nonetheless economically significant. Sunny, dry summer weather yields profits for businesses, with people flocking to campgrounds and beaches—but also in buying swimsuits, and even buying annual gym memberships so they can work out and look good in those swimsuits; this in turn affects the employment of personal trainers and swim instructors. Further back along the supply chain, beach weather creates demand in the fashion industry for the talent of swimsuit design; indeed, the fashion industry times the design and release of all its collections to the change of seasons.

Cold winter weather gets the cash registers ka-chinging for the sellers of heating oil, snow tires, down parkas, and hot chocolate. Further back along the supply chain, winter cold also creates demand for goose down and the cultivation of cacao.

In our global economy, even weather events halfway around the world have cash repercussions. A poor growing season in Africa or South America that yields a shortage of coffee beans or sugar, for example, is felt in the wallet by the coffee-drinking morning commuter in Europe and the United States.

A resort in Mexico beckons Northerners away from wintry weather, and fuels the local economy.

Costs of Weather Disasters

Weather disasters have enormous impact on local, regional, and national economies. According to statistics kept by the National Weather Service, over the past couple of decades, heat-related deaths have far outstripped all other disasters, constituting nearly half of fatalities during the record year of 1995. Over the past 30 years, floods have claimed the most lives. In earlier years, tornadoes and lightning were leaders.

In 2005, hurricanes Katrina and Rita had unprecedented effects. Together, the two hurricanes reduced the United States GDP by 0.7 and 0.5 percentage points in the third and fourth quarters of 2005. Not only were local businesses, schools, homes, and hospitals affected, but also the regional power grid, regional wireline and wireless communications, interstate highways, rail and air transportation, and the delivery of the U.S. mail.

Weather disasters also alter patterns of human migration and settlement. In the months following the two 2005 hurricanes, some three-quarters of a million suddenly homeless Louisianans, Mississippians, and Alabamans set down

Above: NASA technician Roberto Caballero tests the sounder instrument's cooler cover door on the GOES-L weather satellite. The United States invests an extensive amount of time and labor in predicting the weather. Top left: Scientists at the South Pole make adjustments to a background infrared measuring instrument. This tool will help them monitor and even predict changes in the weather.

new roots elsewhere in the country—the single largest mass migration in U.S. history since the dust bowl era of the 1930s.

VALUE OF PREDICTION

Weather prediction is not cheap. It requires costly resources in the form of satellites, massive supercomputers, laboratories, weather radars and other instruments, and a detailed network of weather stations and skilled personnel.

Numerous studies conducted by the National Oceanic and Atmospheric Administration (NOAA) and other organizations and researchers, however, repeatedly demonstrate that the investment is recouped many times over in the property and lives saved by storm predictions. Just in the United States alone, in Florida and along the Gulf Coast, state and local authorities use hurricane warnings to trigger evacuations, saving many lives. Similarly, the number of people killed by tornadoes today is only 10 percent the number of fatalities a century ago, even though population densities are greater—primarily because of tornado warnings, tornado drills in schools, and mandatory tornado shelters in public places. Warnings about severe thunderstorms can allow airlines to reroute aircraft around storms, or fishermen or leisure boaters to get ashore; severe frost warnings alert farmers so they can take measures to protect crops. Flood warnings, blizzard warnings, and predictions of

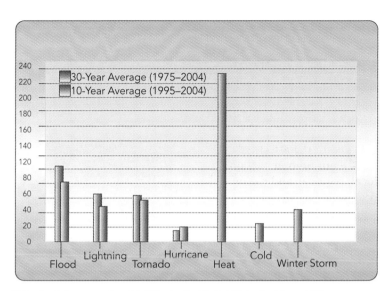

This chart shows weather fatalities in the United States in 10- and 30-year averages. The information was compiled in 2004, before Hurricanes Katrina and Rita, and reveals that mortality from heat-related causes far outstrips any other recent trends.

heat waves or cold snaps have similarly proved their value.

In addition to short-term weather forecasts, long-term seasonal outlooks also have great value, even though the complexity of these systems makes it very difficult to make predictions. Meteorologists and climatologists are now seeking to predict whether or not certain years might have an El Niño or La Niña. Gaining some notion 6 to 9 months in advance of a growing season's expected range of temperatures and levels of

precipitation could prevent losses estimated to be in the billions of dollars.

This value of such predictions would extend outside the United States to farmers in Mexico, wool-producers in Australia, wheatgrowers in Argentina, and other regions whose climates appear to be correlated with the ENSO. Such information would also be useful to electric utilities in allocating load; utility companies have calculated the dollar value of such advance knowledge to be worth millions of dollars per percentage point of accuracy. Moreover, in recent years the life and property insurance industry has been seeking how to factor climate change and the possible effects of the increased number and strength of hurricanes and other weather disasters, and factoring that into insurance risk and premiums for homes and businesses.

Gabriela Pesqueria shows Fanny Scott, a Hurricane Katrina refugee, some options for living in Colorado. Mrs. Scott is among the three-quarters of a million who have relocated due to the devastation caused by the 2005 hurricane.

Weather in War

Weather has always been a crucial factor to warfare. Until the twentieth century's invention of such technologies as infrared night-vision goggles and radar, which allowed the detection of people or aircraft even through clouds or at night, war was primarily fought in the daytime, and preferably also in decent weather, since weather equally assisted or hampered both sides.

One can cite many battles in which weather determined outcome, including decimating the ships of Spain's supposedly invincible Armada in 1588, preventing its invasion of Britain. In three other famous battles—one in the New World, one in Russia, and one in the English Channel—weather was crucial in turning the tide of war.

AMERICAN REVOLUTION, 1776

Bitter winter snows and other bad weather significantly affected the course of battle several times during the American Revolution, perhaps the most famous incident being George Washington's crossing of the Delaware River in 1776 and the ensuing Battle of Trenton.

In December, General Washington had chosen Valley Forge, Pennsylvania, as the winter camp for his remaining 5,000 ragtag farmer-soldiers because of its easy defensibility. Many men had deserted, and

the remainder were demoralized because they had not won a single major battle in 21 months. On Christmas night, Washington inspired them by reading aloud Thomas Paine's just published *American Crisis*, and then

Above: General George Washington is shown crossing the Delaware River in this painting. The unusually cold winter of 1776 allowed the troops, once safely across, to trundle guns on roads that were normally muddy and impassable. Lower left: This map shows the Spanish Armada's route through the English Channel and around Ireland in the summer of 1588. Already battered by a major storm en route to England, the Spanish fleet eventually succumbed to one of the northernmost hurricanes on record. Two dozen ships were wrecked off the coast of Ireland. Top left: Napoleon and his troops cross the snowy Russian territory in E. Messonier's painting. The unusually cold winter of 1812 sickened or killed so many of Napoleon's soldiers that Russia's victory was assured.

ROUTES OF THE ARMADA

✕ Fights in the Channel
✕⤬ Wrecks

divided the men into three groups. His idea was for them to steal across the Delaware into New Jersey in the dead of night and launch surprise dawn attacks on three British encampments.

But that night, a wicked nor'easter swept down, complete with hail, sleet, snow, and other freezing precipitation. It slowed their progress—the 300-yard river crossing alone took nine hours—and the temperatures plunged so far that at least two men froze to death. But also frozen hard were the previously muddy roads, so the troops could trundle guns that otherwise would have bogged down. Washington's men arrived in Trenton to find the German mercenary soldiers hired by the British sound asleep in a drunken holiday stupor. Within 90 minutes, the revolutionaries had captured 1,000 prisoners while losing only four men.

The Allies picked June 6, 1944, as D-Day based on favorable weather predictions. The successful Allied invasion of France depended on clear skies and winds of under Beaufort Force 3.

NAPOLEONIC WARS, 1812

In June 1812, controlling nearly all of Europe and at the peak of his power, Napoleon I amassed half a million troops and invaded Russia, seeking to force Czar Alexander I to submit to a treaty. The Russians followed a policy of "strategic retreat" back toward Moscow. They engaged in no battles, but burned all the farmland as they retreated, leaving the French with no supplies. By the time Napoleon took Moscow on September 14, most of its population had fled, leaving little shelter or supplies.

Napoleon decided to march what was left of his army the 1,500 miles back to France. But he completely underestimated the Russian winter. Moreover, the winter of 1812 was unusually early and bitterly cold. So many soldiers died of cold or disease that Napoleon arrived home with only 10,000 men. The campaign assured Napoleon's downfall and Russia's rise.

WORLD WAR II, 1944

Germany held France, just the Channel's width away from England. General Dwight D. Eisenhower and U.K. Prime Minister Winston Churchill knew that decisive battles had to be fought if the Allies were to defeat Adolf Hitler. By 1942, the two leaders had a plan that seemed impossible: invade France along the 21 miles of sandy beaches on the heavily fortified Normandy coast. To land 156,000 men plus tanks, artillery, and supplies, they had to build floating docks because tides rose and fell more than 20 feet twice daily. Moreover, the coastline was infamous for unpredictable stormy weather, even during the summer, but the invasion depended on winds of under Beaufort force 3 and 70 percent clear skies with at least 3 miles (5 km) visibility. Thus, success was absolutely dependent on accurate weather forecasts.

A D-Day of June 6, 1944, was selected by the Meteorological Office, the U.K.'s national weather service. The original date of June 5 was delayed 24 hours because of predictions of wind direction, clouds, and other unfavorable conditions. But June 6, based on visual weather observations, tracking of winds from so-called pilot balloons, and then-current knowledge of weather fronts, the Met Office gave the word. Predictions were for light breezes from the west or west-northwest with clearing skies, favoring the invasion. And the rest, as they say, is history.

Weather
and Civilization

Above: This painting from a 2nd millennium BCE Theban tomb shows a farming scene. The agreeable climate around the Mediterranean and the predictable flooding of the Nile River allowed the ancient Egyptians to produce a stable agrarian society. Below: Ten different reconstructions of mean temperature changes converge on a general curve of global temperature shifts in the past 1,000 years. The curve reveals the current trend of rising temperatures. Recent reconstructions are shown with red lines; older ones with blue. Top left: The first European settlers in North America suffered shockingly cold weather, which contributed to a high mortality rate among the early Pilgrims. They arrived at the coldest point in the period known today as the Little Age Ice.

A number of authors have made a compelling case that weather and climate have strongly affected the course of civilization, exploration, and human culture. Western civilization arose around the "fertile crescent," where moderate weather and adequate rainfall and annual flooding of the Nile and other rivers restored nutrients to the soils that produced fruitful crops. The Medieval Warm Period, with its lack of floating ice, encouraged the Vikings to venture farther and establish settlements on the shores of Greenland, and Europeans to follow great schools of cod in cutthroat competition for wealth.

Similarly, the Little Ice Age—with its exceptional cold spell—dropped temperatures so low that living conditions and year-round sea ice eventually drove the last European settlers off of Greenland. The very coldest part of the Little Ice Age in the seventeenth and eighteenth centuries—coinciding with the Maunder Minimum with its lack of sunspots—accounted for the bitter winters in Plymouth, Massachusetts, in which so many Pilgrims froze or starved to death, or succumbed to disease.

The late eighteenth and early nineteenth centuries were

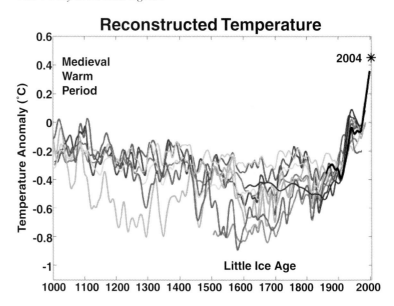

Reconstructed Temperature

Medieval Warm Period

2004 ✳

Little Ice Age

Temperature Anomaly (°C)

also times of extraordinary volcanic activity. Worldwide temperatures in the year or two following a major eruption often show a sharp downward spike, indicating a connection between the eruptions and climate. For example, a powerful series of eruptions of the Icelandic volcano Laki in 1783 and 1784 threw so much ash and sulfates high into the atmosphere that a significant fraction of solar radiation was reflected before it reached Earth's surface. This dimming of the sunlight cooled the northern hemisphere between 1.8° and 5.4°F (1° and 3°C) below normal means. Moreover, because of the peculiarities of atmospheric circulation at high latitudes, the Laki eruption set in motion a series of climatic teleconnections that resulted in weakening the normal monsoon circulations over Africa and India, ultimately reducing the flow of the Nile River and precipitating mass famine.

Again, between 1812 and 1815, there were at least three major volcanic eruptions, this time at lower latitudes; the greatest of them was that of the volcano Tambora in Southeast Asia, causing solar dimming so great that 1816 became known on both sides of the Atlantic as "the year without a summer." In New England, frost formed in July; in Europe, record cold, rain, and snow ruined crops. Resulting famine and hunger triggered the grain riots of 1816 and 1817, whose violence was greater than any social unrest since the French Revolution.

The 1883 eruption of Krakatau was one of the worst volcanic eruptions in recent history. It spewed as much as 6 cubic miles of rock and ash into the atmosphere and spawned 100-foot tsunamis. The amount of material ejected into the troposphere lowered the global temperature by several degrees for almost a decade.

VOLCANO-WEATHER CONNECTIONS

Atmospheric circulation is complex and not fully understood. So the effects of massive volcanic eruptions on climate are also not fully known. But recent research of historical eruptions suggests that low-latitude volcanoes, such as those in the tropics or the Caribbean, may have a different effect on Earth's weather and climate than high-latitude volcanoes, such as those in Alaska or Iceland.

One main difference has to do with the general circulation of the atmosphere, which is preferentially from the Equator toward the poles. Thus volcanic ash and aerosols from a tropical volcano are readily spread poleward in both hemispheres, blanketing the planet. But the ash and aerosols from a high northern-latitude eruption tend to stay closer to the poles, and having different effects in different regions around the globe.

Another factor is the sheer volume of material and the force of the eruption. Not only did Mount Saint Helens in Washington State emit significantly less volume of ash and aerosols than did Laki, Tambora, Krakatau, or the Alaskan volcano Novarupta in 1912, but the vent spewed it out sideways, so most of the material ended up in the lower troposphere. Thus, most of the ash was caught up in rain or other processes that quickly brought it to Earth. The other volcanoes, however, erupted vertically and with such force that much material was shot well into the stratosphere, where it remained suspended for several years.

A third variable is the amount of sulfates compared to the amount of ash. Although particulates of ash can shade Earth from sunlight, they are large enough to fall to Earth in relatively short times (days to months). Sulfates, however, can combine with other chemicals in the atmosphere to create aerosols with different chemical compositions that can stay aloft for several years. It is such aerosols that may reflect sunlight the most and act to cool the planet.

Through a Gas, Darkly?

What do jet contrails and smog have in common with volcanoes? They may be shading Earth or reflecting sunlight, possibly obscuring some of the effects of global warming.

EVIDENCE

The amount of sunlight reaching the ground may be significantly less than it was half a century ago. That is the implication of independent sets of measurements made from the 1950s through the 1990s for different purposes by various scientists around the world. All used standard pan-evaporation experiments, with which one can measure how effectively direct sunlight causes open water to evaporate while controlling for other factors such as ambient heat, humidity, or wind.

Measurements over these four decades revealed that evaporation rates decreased, implying that the amount of sunlight reaching the ground also decreased. The global average was 5 percent, but different regions varied. The amount of dimming over the industrialized United States was closer to 10 percent, whereas over Hong Kong it reached a whopping 37 percent.

Recent work has indicated that slower evaporation rates may be related to increases in atmospheric humidity. Warming temperatures should accelerate the hydrologic cycle, which can increase humidity.

Another important clue came in the three days immediately following September 11, 2001, when four aircraft were used in terrorist attacks. With all civil aircraft downed across the United States, the skies—sunny and cloudless across the entire nation—were also completely clear of crisscrossing jet contrails. When atmospheric scientists examined temperature records for hundreds of locations around the country during those three days, they observed a telling pattern: an increase in the spread of temperatures between daily highs and lows by as much as 1.8°F (1°C).

Above: Hong Kong is one of many cities blanketed by smog. While smog clouds may mitigate the effects of global warming, they pose many other problems to life on Earth. Top left: Jet contrails may mask some of the effects of global warming by producing vapors that reflect the Sun's heat back into space.

EXPLANATION

Aerosols can reflect sunlight, shading and dimming the ground. Aerosols also form condensation nuclei that cause water vapor to condense into cloud droplets. Cloud droplets are mirror-like, reflecting sunlight into space. Together, they cut back on the amount of sunlight—solar radiation—reaching the ground. Less solar radiation means less water evaporating from the experimental pans.

The absence of clouds and jet contrails showed the converse case, and was indeed what might expect if global dimming is real. Clouds not only cool daytime temperatures by reflecting sunlight, they also raise nighttime temperatures by reflecting heat (long-wavelength infrared) radiation from the ground back down to Earth. So an absence of clouds and contrails should result in both higher daytime temperatures and lower nighttime temperatures—exactly what was observed.

IMPLICATIONS

Aerosol pollution and increased cloud cover resulting from warming temperatures can both shade the surface. If global dimming is cooling the planet, is it counteracting global warming?

The answer seems to be both yes and no. More clouds would cool the planet in the daytime by reflecting sunlight. But the type of clouds are not necessarily desirable: clouds formed from water vapor condensing on smog aerosols are basically clouds formed from pollutants, likely giving rise to more acidic precipitation. So the answer to global warming is not to release more smog to form aerosol clouds to cool Earth.

This haze over Northern India consists mainly of pollutants produced by human industry. Aerosol pollution poses health hazards for humans and other life forms, and is largely responsible for the phenomenon of global dimming.

Some climatologists hypothesize, however, that the existence of such aerosol clouds has obscured the true magnitude of global warming. Moreover, some even feel that it is important to get levels of carbon dioxide and other greenhouse gases under control before eliminating smog aerosols, because in the absence of the aerosol clouds, global temperatures might climb very fast. Others question that logic, because other measurements suggest that the dimming may be less global than regional or even local; dimming is greater over latitudes with industrialized nations, and almost nonexistent over unpopulated areas.

The trend toward global dimming seems to have reversed since about 1990, largely because European nations and other developed countries have reduced aerosol emissions.

Human Effects on Climate

Calculate how many gallons of gasoline you pumped last year into your car(s) and any other internal combustion engines in your life. Count the number of "disposable" plastic bottles, foam take-out containers, or plastic sandwich bags or grocery bags you and your family discarded. Figure out how many cubic feet of natural gas, gallons of heating oil, or kilowatts of electricity your house consumes in a year in heating, cooling, and lighting.

Now multiply this for billions of people across the globe. The amount of resources being consumed will expand as the human population continues to grow.

Neither gasoline nor plastics nor even much electricity would exist had hydrocarbons not been extracted from Earth as crude oil or coal. After being deposited in the days of the dinosaurs, that oil and coal resided underground for millions of years. Now, in scarcely a century, vast amounts of it have been removed from the ground and used to power our growing industrial society. The waste products are being released into the atmosphere, in part greenhouse gases such as carbon dioxide and methane. Today the concentration of atmospheric carbon dioxide is higher than it has been in at

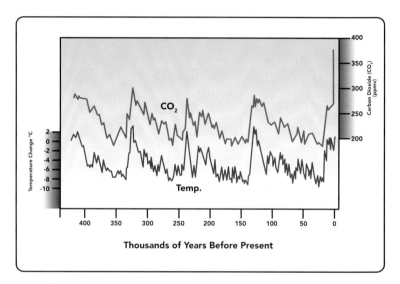

Above: The rise in both global temperature and global levels of carbon dioxide over the last 450 thousand years is shown in this chart, based on the 2001 assessment of the UN Intergovernmental Panel on Climate Change (IPCC). Below: This sign marks the withdrawal of Exit Glacier from its 1978 position. Shrinking glaciers are yet another warning of global warming. Top left: Disposable packaging belies our society's reliance on crude oil and leaves a legacy of plastic that will take half a century to decompose.

least 400,000 years, perhaps even for the entire Holocene era.

Although some individuals were concerned about this trend as far back as the 1960s, scientific knowledge was still uncertain enough over succeeding decades that there was room for disagreement about the significance of the data. But since the 1990s, many signs worldwide have been unmistakable: the rate of shrinkage of both the Greenland ice cap and Antarctic ice; the melting of glaciers in Alaska, the Tyrol, and Kilimanjaro and better understanding of the connection between temperature and atmospheric carbon dioxide.

With the publication of scientific papers and reports by the Intergovernmental Panel on Climate Change (IPCC), the consensus is clear: temperatures are rising, and human activities are having an impact on the process.

Cashmere goats, like the one above, have overgrazed the steppe grasslands of the Alashan Plateau, due to increased American demand for cashmere sweaters. The resulting destruction of the soil integrity is causing dust storms like the one pictured at left in Beijing.

DEMAND + SUPPLY = DESERTIFICATION?

Global trade can affect climate across the planet. One example is the unintended impact of inexpensive cashmere sweaters that flooded American discount stores in the 1990s.

Production of these sweaters skyrocketed in China. That created demand for more cashmere-producing goats. These goats overgrazed the steppe grasslands in the Alashan Plateau down to bare soil, and their hooves crushed tender plant roots. These and many other land-use changes slowly led to the removal of vast amounts of the grass cover from the Chinese and Mongolian prairies. Just between 1994 and 1996, the Gobi Desert expanded in size by an area the size of the Netherlands.

China is now suffering dust storms as a result of desertification. Such dust storms, which regularly engulf Beijing and even North Korea, have more than quadrupled in frequency from perhaps five times a year in the 1950s to nearly every other week today. Several storms have been so huge that Asian dust has even caused health problems for residents of Washington, Oregon, Idaho, and British Columbia, and reduced visibility as far west as Colorado.

GLOSSARY

A

ADVECTION The horizontal transport of atmospheric properties (including heat, moisture, chemicals, turbulence) effected by the mass motion of the atmosphere.

AEROSOLS Fine liquid droplets or solid particles (including fog, haze, smoke) dispersed throughout a gas, such as air, important because they act as nuclei for the condensation of water or the deposition of ice.

AIR MASS A widespread body of air, homogeneous in temperature or moisture, that has taken on the properties of the region of Earth's surface below it.

ATMOSPHERIC OPTICS (METEOROLOGICAL OPTICS) The study of optical phenomena in the atmosphere that are the product of atmospheric properties, such as the interaction of sunlight with water droplets, ice crystals, or dust to create halos, mirages, or rainbows.

AURORA The flickering "northern lights" or "southern lights" occasionally seen at high and middle latitudes at night, originating when magnetic storms on the sun send out charged subatomic particles that interact with oxygen and nitrogen molecules in Earth's upper atmosphere.

B

BAROMETER An instrument that measures the atmospheric pressure of ambient air.

C

CIRRUS (abbreviated Ci) A principal cloud type (genus) that occurs at high levels in the troposphere composed primarily of ice particles, and appearing as wispy white streaks or delicate white filaments or patches.

CLIMATE The complete global system of atmospheric, terrestrial, and oceanic conditions over a long period of time (years or longer); also, the average pattern of aggregate weather conditions and variations characteristic of a given locale over a long period of time.

CLOUD BASE The lowest level in the atmosphere with a perceptible number of cloud particles, either water droplets or ice crystals; for certain clouds, the cloud base is visibly flat at one distinct altitude.

CONDENSATION A physical process by which water vapor (individual water molecules) transforms into liquid water in dew, fog, or cloud; the opposite of evaporation.

CONDUCTION The transport of energy, such as heat, solely as a result of the random motions of individual molecules, such as by temperature gradients.

CONVECTION Turbulent mass motions within a fluid that result in the mixing of that fluid and its properties (heat, moisture); in the atmosphere, convection is the dominant means of vertical energy transport.

C

CONVECTION CELL An organized unit of convection within a convection layer; examples range from individual thunderstorm cells (with a central updraft and rain and downdrafts at the periphery) to the major Hadley, Ferrel, and Polar cells that transport solar energy from Earth's equator to poles.

CORIOLIS FORCE A force on the atmosphere resulting from Earth's rotation, deflecting air particles to the right (clockwise) in the northern hemisphere and to the left (counterclockwise) in the southern hemisphere, and giving rise to the cyclonic circulation of hurricanes and other weather systems.

CUMULUS (abbreviated Cu) A principal cloud type (genus) that can occur at any level in the troposphere, consisting of large, detached elements of rising domes or heaps whose appearance somewhat resembles cauliflower.

CYCLE A process in which the ending state is the same as the beginning state; cycles may be open (such as Earth's energy cycle, because the planet is always receiving energy from the sun and radiating it back into space) or closed (such as the water, carbon, or oxygen cycle, in which atoms or molecules are endlessly recycled).

D

DEPOSITION A physical process by which water vapor transforms into solid ice particles without going through an intervening liquid phase; the opposite of sublimation.

DEW Water condensing onto leaves, grass, or other objects as a result of their temperatures falling below the dew point (a temperature above freezing at which dew forms).

DIFFRACTION An optical "edge effect" in which light is spread out when it passes through an aperture or grazes a particle whose diameter is close to the wavelength of the light; important in meteorological optics.

E

EL NIÑO A significant warming in the sea surface temperatures of the eastern Pacific Ocean near the equator that recurs irregularly roughly every two to seven years.

EL NIÑO–SOUTHERN OSCILLATION (ENSO) A term recognizing the linkage of the oceanic El Niño warming in the eastern Pacific with a see-sawing Southern Oscillation pattern in sea level atmospheric pressure on both sides of the Pacific; ENSO is also associated with other anomalies in temperature, precipitation, and air circulation elsewhere around the globe.

EVAPORATION A physical process by which a liquid is transformed to its gaseous state; the opposite of condensation.

EXOSPHERE The outermost layer of Earth's atmosphere, the only layer from which molecules (mostly traces of helium and other atoms from the sun) can escape into outer space; even so-called airless bodies, such as the Moon or Mercury, have an exosphere.

F

FERREL CELL A global atmospheric circulation pattern in the troposphere, also known as the zone of mixing, between the Hadley Cell and Polar Cell, involved in the transport of warmth and moisture from equator to poles.

FRONT (COLD, WARM, STATIONARY, OCCLUDED) An interface or boundary between two air masses of different densities, usually also of different temperatures and humidities; cold and warm fronts are the leading boundaries of masses of cold and warm air, a stationary front is one that exists between two air surface masses that are essentially stalled, and an occluded front also may involve an air mass that is forced aloft.

FROST Ice crystals formed by deposition onto objects as a result of their temperatures falling below the frost point (the name often used when the dew point is below freezing).

EQUINOX The first days of spring and autumn on or about March 21 and September 22 when the sun crosses the celestial equator.

G

GRAUPEL Small snow pellets, also known as "soft hail" or "tapioca snow" for their appearance, formed from heavily rimed snowflakes until they are approximately spherical or conical and about 2–5 mm across.

GREENHOUSE EFFECT The heating effect of the atmosphere on Earth because of trace gases in the atmosphere absorbing and reemitting long-wavelength solar infrared radiation toward the ground instead of letting it escape into space.

GREENHOUSE GAS Trace gases in Earth's atmosphere that give rise to the greenhouse effect, notably carbon dioxide, methane, water vapor.

H

HADLEY CELL An atmospheric circulation pattern in the troposphere driven both by convection and by Earth's rotation, consisting of rising warm air near Earth's equator and sinking air in the subtropics.

HAIL Balls or lumps of ice ranging in size from peas to softballs precipitating from convective clouds, usually cumulonimbus (thunderheads).

HUMIDITY Some measure of water vapor in the air, either absolute or relative (see pages 26–27 for fuller description).

HYDROLOGIC (WATER) CYCLE Ongoing processes in which water evaporates from oceans and land, is carried around the globe, precipitates, runs off land, recharges aquifers, and eventually ends up back in the oceans again to repeat the cycle.

HYGROMETER Any instrument for measuring atmospheric humidity.

I

INFRARED Electromagnetic radiation longer than red wavelengths of light and shorter than microwaves that includes thermal (heat) radiation.

IONOSPHERE An atmospheric region containing significant ions and electrons, collocated with the mesosphere and thermosphere and extending hundreds of kilometers outward into space around Earth;

important less for weather than for the transmission of radio signals.

IONIZATION Any of several physical processes that can strip neutral atoms or molecules of outer electrons, leaving negatively charged electrons and positively charged ions.

J

JET STREAM A quasi-horizontal narrow current of high-speed winds in the upper troposphere, commonly found at the boundary of the Ferrel and Polar cells (the polar jet) and sometimes also at the boundary of the Ferrel and Hadley cells (the subtropical jet).

L

LA NIÑA A significant cooling in the sea surface temperatures of the eastern Pacific Ocean near the equator that recurs irregularly roughly every two to seven years; the cold phase of ENSO.

LATENT HEAT The amount of energy absorbed or released when a substance changes phase; latent heat is absorbed when ice melts or water evaporates, and is released when water vapor condenses into water droplets or ice crystals; latent heat is important in the transfer of solar energy from Earth's equator to poles.

M

MAUNDER MINIMUM The prolonged period from 1645 to 1715 in which sunspots were virtually absent from the surface of the Sun, coinciding with the coldest years of the Little Ice Age that stretched roughly from the fourteenth through the eighteenth centuries.

MESOSPHERE The middle and coldest layer of the atmosphere, located between the stratosphere and the thermosphere, and including part of the ionosphere.

MONSOON A name for prevailing winds that switch with the seasons, and that usually usher in a wet season alternating with a dry season.

N

NIMBUS A former name used for any rain-producing cloud, but now used as a suffix or prefix to describe rain-producing clouds of different forms (nimbostratus, cumulonimbus, etc.).

O

OCEAN CONVEYOR BELT (THERMOHALINE CIRCULATION) The three-dimensional global circulation of the oceans' water masses that determines today's climate, whose motions are driven by solar heating at the equator and freshwater near the surface from precipitation, largely responsible for much of the heat transport by the oceans.

OROGRAPHIC The effects of mountains or mountain ranges on weather, such as wind flow, clouds, temperature, humidity, and precipitation.

OZONE A nearly colorless (slightly bluish) gas composed of molecules consisting of three oxygen atoms, commonly created in the lower atmosphere when lightning ionizes air; ozone strongly absorbs ultraviolet radiation, and a layer of ozone in the stratosphere plays an important role in shielding Earth from damaging solar UV radiation.

P

POLAR CELL A global convection cell in the troposphere characterized by rising air at high latitudes descending air over Earth's poles.

PRESSURE, ATMOSPHERIC The pressure or weight exerted by a vertical column of air directly over the instrument resulting from the downward pull of gravity on the atmosphere; measured by barometers, it is also called barometric pressure, and varies widely around the globe depending on weather systems.

PSYCHROMETER A device for measuring the dew point and humidity by placing two thermometers next to each other, one with a dry bulb and one with a wet bulb, and measuring the respective rates of evaporation.

R

RADIATION The physical process of electromagnetic radiation (light, heat) being conveyed through free space, such as from the Sun to Earth.

RAIN Precipitation in the form of water droplets heavy enough to fall.

REFLECTION A change of direction in the travel of a light wave as it bounces off an external or internal surface of a medium, such as raindrops, snow, clouds, or ice crystals; reflection may also change the light wave's amplitude (brightness) if some light is absorbed (as when light reflects off the ocean).

REFRACTION A change not only in direction and amplitude of a light wave, but also in wavelength of a light wave as it travels through a medium, such as an ice crystal or raindrop, resulting in white light being spread out into a rainbow spectrum of colors.

RESERVOIR A space capable of storing a substance, not only in the common sense of a lake being a reservoir for water, but also in the sense of trees being a reservoir for atmospheric carbon.

RIME A milky white deposit of ice crystals formed by the rapid freezing of supercooled water drops hitting a cold object; differs from frost, which is formed by the direct deposition of water vapor molecules from air.

SCATTERING (LIGHT) A change in the direction of travel of a light wave in all directions, including at right angles, as it passes through aerosols or another medium; sunbeams are visible because sunlight is scattered from dust particles in air.

STRATOSPHERE A vertically stable (calm) layer of the atmosphere above the troposphere and below the mesosphere, where temperature increases with height; it is also the location of the ozone layer.

STRATUS (abbreviated St) A principal cloud type (genus) in the form of a gray layer of a relatively uniform cloud base.

SUBLIMATION A physical process by which ice transforms into water vapor without going through an intervening liquid phase; the opposite of deposition.

SUNSPOT A magnetic storm on the surface of the sun, often larger than the planet Earth, that looks relatively dark; the number of sunspots waxes and wanes in a cycle of roughly 11 years, and their presence is associated with greater solar magnetic activity and the ejection of solar gases throughout the solar system, giving rise to auroras on Earth.

SUPERCOOLING The chilling of water (or another liquid) below its normal freezing temperature without having it become a solid, usually because of a lack of aerosols that can serve as nuclei for the crystallization of ice; supercooling of water droplets in clouds is very common, and is important for many weather processes, including the formation of snow and freezing rain.

SYNOPTIC The gathering of comprehensive meteorological data from simultaneous measurements over a wide area (hundreds of kilometers across), and the presentation of that information in a large-scale overall "snapshot" of the state of the atmosphere for that area.

TEMPERATURE Thermal energy as measured by a thermometer; not identical to heat or cold, although colloquially the terms are often incorrectly used as synonyms (see pages 20–21 for detailed explanation).

THERMOMETER An instrument for measuring temperature.

THERMOSPHERE The layer of the atmosphere between the mesosphere and exosphere, and including part of the ionosphere.

TRADE WINDS Prevailing global subtropical and tropical wind systems, blowing from the northeast (toward the west) in the northern hemisphere and from the southeast (toward the west) in the southern hemisphere.

TROPOSPHERE The lowest layer of the atmosphere down to the ground, the one in which most moisture resides and most weather occurs.

ULTRAVIOLET Energetic electromagnetic radiation shorter than violet wavelengths of light and longer than X rays; fully 9 percent of solar radiation is UV, which is responsible for complex and important interactions with Earth's atmosphere, including the formation of the ozone layer in the stratosphere.

WALKER CELL A pattern of air circulation and atmospheric pressure spanning the Pacific Ocean near the equator, whose variations—called the Southern Oscillation—are now known to be associated with the periodic upwelling of warm water in the eastern Pacific known as El Niño.

WEATHER The short-term state of the atmosphere (temperature, humidity, pressure, winds, cloudiness, precipitation) on a scale of minutes to days.

FURTHER READING

BOOKS

Aguado, Edward, and James E. Burt. *Understanding Weather and Climate*. Upper Saddle River, N.J.: Pearson Prentice Hall, fourth edition, 2007.

Bluestein, Howard B. *Tornado Alley: Monster Storms of the Great Plains*. New York: Oxford University Press, 1999.

Burroughs, William J., Bob Crowder, Ted Robertson, Eleanor Vallier-Talbot, and Richard Whitaker. *A Guide to Weather*. San Francisco: Fog City Press, 1996 (latest edition 2005).

Burt, Christopher C. *Extreme Weather: A Guide and Record Book*. New York: W.W. Norton Co., 2004.

Egan, Timothy. *The Worst Hard Time: The Untold Story of Those Who Survived the Great American Dust Bowl*. New York: Houghton Mifflin, 2006.

Greenler, Robert. *Rainbows, Halos, and Glories*. New York: Cambridge University Press, 1980.

Huler, Scott. *Defining the Wind: The Beaufort Scale, and How an 19th-Century Admiral Turned Science into Poetry*. New York: Three Rivers Press, 2004.

Intergovernmental Panel on Climate Change, *IPCC Third Assessment Report—Climate Change 2001* (fourth assessment report due out in 2007; other reports on the ozone layer, carbon dioxide, and other climate-related subjects also available), www.ipcc.ch/pub/online.htm

Lamb, H. H. *Climate, History and the Modern World*. London: Routledge, 1982, reprinted 1997.

Laskin, David. *The Children's Blizzard* [blizzard of 1888]. New York: Harper Perennial, 2004.

Ludlum, David M. *National Audubon Society Field Guide to Weather*. New York: Alfred A. Knopf, 1991 (eleventh printing, 2002).

Meinel, Aden, and Marjorie Meinel. *Sunsets, Twilights, and Evening Skies*. Cambridge: Cambridge University Press, 1983.

Minnaert, M.G.J. *Light and Color in the Outdoors*. New York: Springer, 1992 (the latest in a whole series of reprints of a classic early twentieth-century book).

Monmonier, Mark. *Air Apparent: How Meteorologists Learned to Map, Predict, and Dramatize Weather*. Chicago: University of Chicago Press, 1999.

Naylor, John. *Out of the Blue: A 24-hour Skywatcher's Guide*. Cambridge: Cambridge University Press, 2002.

Ochoa, George, Jennifer Hoffman, and Tina Tin. *Climate: The Force That Shapes Our World—and the Future of Life on Earth*. London: Rodale International Ltd., 2005.

Oliver, John E., ed. *Encyclopedia of World Climatology*. Dordrecht, Netherlands: Springer, 2005.

Rauber, Robert M., John E. Walsh, and Donna J. Charlevoix. *Severe and Hazardous Weather*. Dubuque, Iowa.: Kendall/Hunt, 2002.

Schmidlin, Thomas W., and Jeanne Appelhans Schmidlin. *Thunder in the Heartland: A Chronicle of Outstanding Weather Events in Ohio*. Kent, Ohio: Kent State University Press, 1996.

Schneider, Stephen H., ed. *Encyclopedia of Weather and Climate*. New York: Oxford University Press, 1996.

University of Washington, *El Niño and Climate Prediction*, Reports to the Nation on Our Changing Planet, www.atmos.washington.edu/gcg/RTN/rtnt.html

Williams, Jack. *The Weather Book*. New York: Vintage Books, 1992.

WEB SITES

American Meteorological Society. *Glossary of Meteorology.* Allen Press, 2000 | amsglossary.allenpress.com/glossary

Arctic Climate Impact Assessment | amap.no/acia

Arctic Monitoring and Assessment Programme www.amap.no

British Meteorological Office guide to cloud types www.metoffice.gov.uk/publications/clouds/index.html

Intergovernmental Panel on Climate Change www.ipcc.ch (established by the World Meteorological Organization and the United Nations Environment Programme; see www.ipcc.ch/about/about.htm for background)

JetStream: the U.S. National Weather Service Online School for Weather | www.srh.noaa.gov/jetstream

National Climatic Data Center www.ncdc.noaa.gov/oa/ncdc.html

National Hurricane Center | www.nhc.noaa.gov

National Severe Storms Laboratory, Norman, Oklahoma www.nssl.noaa.gov

National Snow and Ice Data Center www.nsidc.colorado.edu

"Ozone Depletion," U.S. Environmental Protection Agency | www.epa.gov/ozone

Pew Center on Global Climate Change www.pewclimate.org

SnowCrystals.com—gorgeous photography of snowflakes and a good bit of chilly science www.its.caltech.edu/~atomic/snowcrystals

U.S. Army Corps of Engineers, Cold Regions Research and Engineering Laboratory | www.crrel.usace.army.mil

Build your own backyard weather station from household items; instructions appear at several sites:

Center for Innovation in Engineering and Science Education | www.k12science.org/curriculum/weatherproj2/en/activity1.shtml

DiscoverySchool.com | school.discovery.com/lessonplans/activities/weatherstation/airwaterboth.html

Franklin Institute | www.fi.edu/weather/todo/todo.html

Miami Museum of Science www.miamisci.org/hurricane/weatherstation.html

MAGAZINES

Weather. Royal Meteorology Society | www.rmets.org

NOAA Magazine. National Oceanic and Atmospheric Research | www.magazine.noaa.gov

Weatherwise. Heldref Publications | www.weatherwise.org

ORGANIZATIONS

American Meteorological Society | www.ametsoc.org

European Meteorological Society | www.emetsoc.org

The Royal Meteorological Society | www.rmets.org

World Meteorological Organization | www.wmo.ch

AT THE SMITHSONIAN

The Smithsonian Institution in Washington, D.C., the world's largest museum and research complex, is composed of nineteen museums plus the National Zoo. Twenty-four million visitors a year visit the Smithsonian exhibitions that display only a fraction of its vast collections of more than 142 million objects. While best known for its museums, the Smithsonian also runs an impressive network of laboratories, long-term research sites, and field sites from Alaska to Antarctica. These facilities provide a wealth of opportunities for researchers and students interested in global change and environmental biology, with a particular focus on atmospheric conditions, astronomy, ecosystem dynamics, and how weather and climate impact society. The Smithsonian played a central role in early understanding of weather patterns in the United States. In the 1850s, the Smithsonian collected weather reports from hundreds of telegraph stations. This observation network led to the creation of the National Weather Bureau.

THE NATIONAL AIR AND SPACE MUSEUM (NASM)

www.nasm.si.edu

The National Air and Space Museum houses a vast collection of aircraft, including a prototype of the TIROS-1 (Television and Infra-Red Observation Satellite), the first weather satellite. Launched in 1960, TIROS-1 amazed both forecasters and scientists with breathtaking photographs of the Earth's weather systems. The satellite transmitted thousands of images of cloud patterns and other phenomena to ground stations during the life of its three-month orbit. Its modern descendants make space-based weather observations a commonplace component of television newscasts.

The Smithsonian's National Air and Space Museum on the National Mall, Washington, D.C.

Above: The Smithsonian Castle silhouetted in a November sunrise. Left: TIROS-1 (Television Infrared Observation Satellite), the first successful weather satellite, was launched on April 1, 1960. It is on display at the National Air and Space Museum.

SMITHSONIAN ENVIRONMENTAL RESEARCH CENTER (SERC)

www.serc.si.edu

The Smithsonian Environmental Research Center (SERC) is the world's leading research center for environmental studies of the coastal zone. At this 2,800-acre facility on the shores of Chesapeake Bay and the Rhode River, the scientists pursue a wide array of questions related to coastal ecosystems. One major focus is the effect of human-induced environmental changes, which include pollution, climate change, and the spread of non-native species. Because these changes occur in concert on land and in marine waters, researchers are investigating how change in one area of an ecosystem affects the other elements of that ecosystem. The Center's terrain includes forest, cropland, pasture, freshwater wetlands, tidal marshes, and estuaries. Facilities include: an instrument tower, forest canopy tower, meteorology station, carbon dioxide chambers, greenhouse, lath house and experimental gardens, fish weir, and water quality laboratory. Outreach and education are a large component of SERC's mission, allowing the public and students access to both hands-on and distance-learning opportunities.

The SERC Branch of Smithsonian Institution Libraries supports education and research in both the laboratory and the field with collections of books and journals on the topics of global change, environmental science, landscape ecology, and environmental issues. There is even a separate collection known as "Chesapeakiana," relating to the Chesapeake Bay and its environs.

SMITHSONIAN TROPICAL RESEARCH INSTITUTE (STRI)

www.stri.org

Some three million years ago, the Isthmus of Panama rose from the sea, forming a barrier between the Atlantic and Pacific oceans. This event had profound effects on the Earth's climate and the evolutionary and ecological trajectory of marine communities in the region. With field stations on both sides of the isthmus, STRI is the perfect place for investigating the differences between the two oceans and the changes that have occurred.

Shoreline of Wye Island on the Chesapeake Bay. The Smithsonian Environmental Research Center is located 25 miles from Washington, D.C., along the western shores of Chesapeake Bay, which is the largest estuary in the United States. Coastal environments are the focus of the Center's research.

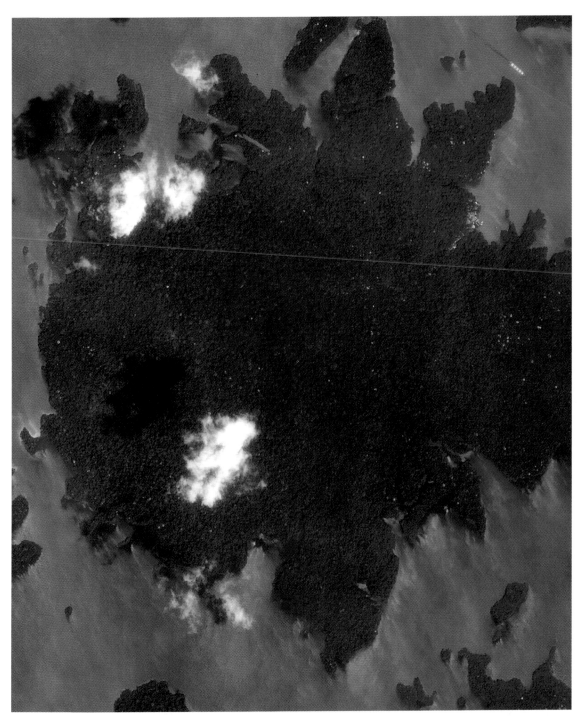

Barro Colorado Island is located in Gatun Lake, at the northern end of the Panama Canal. The red roofs of the Smithsonian Tropical Research Institute can be seen peeking through the dark green canopy inland of the bay on the northeast shore (top right in this photo). Barro Colorado Island has been managed by the Smithsonian since 1924 and is one of the premier sites in the world for the study of tropical forests and the plants and animals that live there.

INDEX

ACKNOWLEDGMENTS & PICTURE CREDITS

Every book is a product of many hands; this one is no exception. On the project itself, thanks go to Aaron Murray (Hylas) for opportunity, Elizabeth Mechem (Hylas) for editorial skill and bantering pleasantries, and advisor Tom Schmidlin (Kent State University) for his factual hawk-eye and skepticism. For leads of facts and photos, thanks go to Les Cowley (owner of the spectacular and instructive site www. atoptics.co.uk/phen800.htm), Dave Dooling (National Solar Observatory), Karl Esch (freelance science photojournalist in Portland, Oregon), Jerry Schad (Mesa College), Dennis Swinford (McKinsey & Co.), and Craig B. Waff (Andrews Air Force Base). Great gratitude is also due to scientists who consented to be interviewed, plus innumerable book, article, and website authors about meteorology terrestrial and extraterrestrial. For being there to be written about, I thank Earth with its atmosphere and weather—which I now appreciate significantly more after my involvement with this book. For forbearance during too many late nights with only cold sandwiches, but happily immersed in my own passion for theater, my warmest thanks and love go to my daughter Roxana (and her quadruped Garrison), to whom this book is dedicated.

The author and publisher wish to thank consultant Thomas W. Schmidlin, Kent State University; Andrew Johnson, Center for Earth and Planetary Studies, National Air and Space Museum; Ellen Nanney, Senior Brand Manager with Smithsonian Business Ventures; Katie Mann and Carolyn Gleason with Smithsonian Business Ventures; Collins Reference executive editor Donna Sanzone, editor Lisa Hacken, and editorial assistant Stephanie Meyers; Hydra Publishing president Sean Moore, publishing director Karen Prince, senior editor Elizabeth Mechem, editorial director Aaron Murray, art director Brian MacMullen, designers Erika Lubowicki, Ken Crossland, Eunho Lee, Pleum Chenaphun; editors Sylke Jackson, Marcel Brousseau, Ward Calhoun, Suzanne Lander, Rachael Lanicci, Michael Smith, and Amber Rose, picture researcher Ben DeWalt, indexer Jessie Shiers; Wendy Glassmire of the National Geographic Society; Harriet Mendlowitz of Photo Researchers, Inc.; and Chris, Sophia, and Linda Carroll.

PICTURE CREDITS

The following abbreviations are used: IS–Istockphoto. com IO–IndexOpen.com; BS–Bigstockphoto.com; SS–Shutterstock.com; JI–Jupiter Images; AP–Associated Press NOAA–National Oceanic and Atmospheric Administration; NGS–National Geographic Society; NWS–National Weather Service; MSFC–Marshall Space Flight Center; WORF–Window Observational Research Facility; CBW–Cambridge Bay Weather; USWB–United States Weather Bureau; GSFC–Goddard Space Flight Center; NMNH–National Museum of Natural History; GWAP–Global Warming Art Project; GISS–Goddard Institute for Space Studies OAR–Oceanic and Atmospheric Research; ERL–Environmental Research Laboratories; NSSL–National Severe Storms Laboratory; USDC–United States Department of Commerce; USAF–United States Air Force; LoC–Library of Congress; MSSS–Malin Space Science Systems; JPL–Jet Propulsion Laboratory; LPI–Lunar and Planetary Institute; KSC–Kennedy Space Center; ESA–European Space Agency EO–Earth Observatory; USDA–United States Department of Agriculture; BARC–Beltsville Agricultural Research Center; NHC–NASA Health Council; JTWC–Joint Typhoon Warning Center

Introduction: Welcome to Weather
v NGS/Klaus Nigge vi PR/Keith Kent 1t NGS/Richard Olsenius 1b NGS/Eichard Olsneius 2 IS/Hector Mandel 3t NGS/Norbert Rosing 3b NGS/Maria Stenzel

Chapter 1: An Ocean of Air
4 SS/Nick Stubbs 5t SS/Marja-Kristina Akinsha 5b SS/Mark Atkins 6tl SS/William Attard McCarthy 6br NASA/MSCF 7tl SS/Painted Lens 7r SS/Andrejs Zavadaskis 8tl SS/Jonathan Larsen 8b Pleum Chenaphun 9 NASA 10tl NASA/WORF 10b Photos.com/JI 11 Pleum Chenaphun 12tl SS/Anna Galejeva 12b NASA 13tl SS/Carl Jani 14tl SS/Igor Talpalatski 14bl SS/Bataleur 14r AP/Andy Newman 15tr SS/Tony Strong 15bl Public Domain 16t SS/Photomediacom 16r Public Domain 17tr AP/John McConnico 17b NGS/Ralph Lee Hopkins

Chapter 2: Meteorology's Origins
18 Public Domain 19t BS/Juan E 20tl 19b NGS 20tl SS/Trevor Allen 20bl Clipart 20r Wikipedia/Fenners

21 Pleum Chenaphun 22tl Wikipedia/David R. Ingram 22l Wikipedia/Cyclopaedia 22tr Wikipedia/Daderot 23 AP/Pat Roque 24tl SS/Scott Rothstein 24t NOAA/Sean Linehan 24b Public Domain 25t SS/T.W. 25bl NOAA 25br REDRAW 26d SS/Bulgar Wladimir 26t Wikipedia/CBW 26b SS/Magri 27bl IS/Deniel Tero 27r SS/Jakob Metzger 28l NOAA/Mel Nordquist 28l NOAA 28tr NOAA 28br NOAA 29 NOAA/USWB 30tl NOAA 30l NOAA 30tr NASA 31 SS/Taolmor 32tl NGS/Bo Brannhage 32bl NOAA 32br Wikimedia/Jacobst 33 Public Domain

Chapter 3: Our Connection to the Cosmos
34 NASA/GSFC 35t NASA/JPL 35b SS/Raymond C. Truelove 36t NASA 36bl NASA 36tr Public Domain 37 Pleum Chenaphun 38tl NASA 29B Pleum Chenaphun 39 IS/Brian McEntire 40tl BS/Smokovski Skopje 40br SS/Mares Lucian 40r Pleum Chenaphun 41t NASA 41b NGS/Bill Hatcher 42tl SS/Vasconcelos 42tr NASA 42br PR/Mark Garlick 43 NGS/Beverly Joubert 44tl NASA/JPL 44bl Pleum Chenaphun 44r SS/Kaulitzki 45 NASA/JPL 46tl NASA 46r NASA 46bl SI/Alfred F. Harrell 47 NASA

Chapter 4: Weather vs. Climate
48 SS/Derek Fitzer 49t SS/Gabriel Openshaw 49b IS/Karen Morrill-McClure 50tl IS/Michel de Nijs 50r NGS/Raymond Gehman 51tl SS/Carsten Peter 51tr NGS/Paul Nicklen 51cl NGS/Sisse Brimberg 51cr NGS/Steve Winter 51bl NGS/Phil Schermeister 51br NGS/Norbert Rosing 52tl IS/Juan Carlos Pires 52tbl SS/Iamanewbee 52bl SS/Jason McCartney 53tl NASA 53tr NASA 53b AP/Shahrzad Elghanayan 54tl SS/ANP 54B PR/Andrew Syred 55tr SS/Nik Niklz 55b Pleum Chenaphun 56tl SS/Marcus Brown 56b AP/Rajesh Kumar Singh 57 Pleum Chenaphun 58tl SS/Bychkov Kirill Alexandrovich 58r IS/Nicola Stratford 58b SPL 59t PR/Ed Adams/Montana State University 29b NASA/GISS

Chapter 5: Great Atmospheric Cycles
60 NGS/O.Louis Mazzatenta 61t SS/Keith Levit 61b Photos.com/JI 62tl SS/Galyna Andrushko 62b NGS/Raymond Gehman 63t Wikipedia/Dante Alighieri 63b NMNH 64tl IS/Dave White 64b USGS 65tr PR/Andrew Syred 65b NGS/Sisse Brimberg 66tl SS/Robert Kohlhuber 66b GSFC/NASA 67tr SS/Lars Christensen 67b photos.com/JI 68tl SS/Michael Rosa 68tr Ken Crossland 68br NGS/Darlyne A. Murawski 69 NASA/GSFC 70tl SS/Ingvar Tjostheim 70b IS Ken Crossland 71t SS/Maxime VIGE 71b Photos.com/JI 72tl SS/Ian Bracegirdle 72bl Photos.com/JI 73t NGS/Annie Griffiths Belt 73b NGS/James P. Blair

Chapter 6: How Weather Works
74 NOAA 75t Photos.com/JI 75b Photos.com/JI 76tl SS/Coverstock 76b Ken Crossland 77t PR/Gordon Garradd 77b IS/Ian Johnson 78tl NWS 78b PR/Jack Fields 79t NASA 79b IS/Andrew Penner 80tl NOAA 80bl NOAA/OAR/ERL/NSSL 81t NASA/JPL 81b IS/Mlenny 82tl IS Alexander Kolomietz 82r NOAA 82l IS/Hilary Brodey 83tr RST 83b AP 84tl IO/DesignPics Inc. 84b Ken Crossland 85tr AP/Bullit Marquez 85b NGS/Casten Peter 86tl SS/Stanislav Khrapov 86b NASA/NHC/JTWC/Gary Padgett 87t NASA/Jesse Allen/MODIS Rapid Response Team 87b NOAA/NWS/OPC 88tl AP/Nick Ut 88c Ken Crossland 88b NGS/Michael Lewis 89 Wikipedia/Surrealspaces 90tl NGS/H. Takeuchi 90tr NOAA/Gerhard Schott 90br SS/Pichugin Dmitry 91tl AP/Divyakant Solanki 91b Wikipedia 92tl NOAA 92bl AP/Siddharth Darshan Kumar 92b PR/Chris Sattlberger 93 NOAA/NWS

Ready Reference:
94l PR/Julian Baum 94r LoC 95t SPL/PR 95b Public Domain 96l NOAA/USWB 96r LoC/Arthur Rothstein 97l NASA 97tr AP/Osamu Honda 97b AP/John Brazemore 98t NGS/Peter Carsten 98bl PR/Jim Reed 98br PR/Jim Reed 99tr PR/ Chris Sattlberger 99b PR/Chris Sattleberger 100 NASA/GSFC/PR 102tl AP/Clifford Grabhorn 102tr AP/Clifford Grabhorn 102b NGS/Paul Nicklen 103t NASA 103b Ken Crossland 104 NASA/PR 105t PR/NASA 106b Ken Crossland 106r NGS/Annie Griffiths Belt 107l PR/NASA 107b USDC/NOAA 108t NGS/Peter Carsten 108bl NGS/Peter Carsten 108br Ken Crossland 109t Ken Crossland 109bl PR/Geospace 109br PR/Susan McCartney

Chapter 7: Clouds
110 Photos.com/JI 111 NASA/Liam Gumley SSEW/University of Wisconsin-Madison 111b SS/Eric Gevaert 112tl Photos.com/JI 112b IS/Amy Kimball 113tl IS/Irina Efremova 113t IS/Victor Melniciuc 114l SS/Trout55 114r Science Museum 114b Trudy E. Bell 115 Ken Crossland 116tl IO/AbleStock 116b Photos.com/JI 117t PR/Magrath/Folsom 117b NSG/Carl Heilman II 118tl

IS/Frances Twitty 118tr NOAA/NWS Collection 118br NGS/George Grall 119tl IO/FogStock LLC 119r Trudy E. Bell 120tl USAF 120r Trudy E. Bell 120bl PR/Pekka Parviainen 121 PR/David Hay Jones

Chapter 8: Precipitation
122 NGS/Stephen Alvarez 123t NGS/Raymond Gehman 123b PR/Ken Thomas 124tl NGS/James P. Blair 124r Pleum Chenaphun 124bl Photos.com/JI 125 NGS/Peter Krogh 126tl Johann Schneider 126b NGS/Raymond Gehman 127t SS/Alexander O. Omelko 127b Trudy E. Bell 128tl Wikipedia/Alex Buirds 128r NGS/Tom Murphy 128br PR/Stephen Dalton 129 PR/Gordon Garradd 130tl Trudy E. Bell 130r AP/Charles Miller 130nl Wikipedia/Barfooz 131t NOAA/NWS 131b NOAA/NWS/NOS/Sean Linehan 132tl NGS/Stephen Alvarez 132br SS/Brykaylo Yuriy 133c Ken Crossland 133tl USDA/BARC 133t NOAA/NWS 133tr NOAA/NWS 133bl USDA/BARC 133br USDA/BARC 134tl Wikipedia/Soon Chun Siong 134bl PR/Kenneth Libbrecht 134r USDA/BARC 134r USDA/BARC 134br USDA/BARC 135 NGS/Soon Chun Siong 136tl NSG/Paul Nicklen 136t NGS/Gordon Wiltsie 136b NGS/Sarah Leen 137tl NGS/Richard Olsenius 137br FogQuest/Tony Makepeace 138tl NGS/Dean Conger 138r Ken Crossland 139tl NGS/Gordon Wiltsie 139b NGS/Paul Nicklen 139b NGS/Karen Kasmauski

Chapter 9: Extreme, Unusual, and Violent Weather
140 PR/Kent Wood 141t AP/Dave Martin 141b PR/Reed Timmer and Jim Bishop/Jim Reed Photography 142tl NGS/Peter Carsten 142r NGS/Gordon Wiltsie 143tr NGS/Maria Stenzel 143b Ken Crossland 144tl PR/Kent Wood 144b PR/Pekka Parviainen 145t NGS/James L. Amos 145b PR/Jim Reed 146tl NGS/William Albert Allard 146b AP/K.N. Choudary 147t PR/Eric Nguyen 147b Pleum Chenaphun 148tl PR/NOAA 148b Pleum Chenaphun 149t NOAA/NWS 149b PR/NASA 150tl PR/Simon Fraser 150r Pleum Chenaphun 151tl NOAA 151bl AP 152tl AP/Adam Butler 152tr NOAA 152br AP/Antonio Lima 153 AP/Eyal Warshavsky 154tl PR/NOAA 154b AP/Eyal Warshavsky 155t Pleum Chenaphun 155b Pleum Chenaphun

Chapter 10: Atmospheric Displays
156 NGS/Norbert Rosing 157t NGS/Maria Stenzel 157b SS/Roman Krochuck 158tl IS/AVTG 158b PR/Alfred Pasieka 159t Ken Crossland 159B IS 160tl Photos.com/JI 160b NGS/Stephen St. John 161t PR/SPL 161b PR/Pekka Parviainen 162tl SS/Wolf Design 162b NGS/Jodi Cobb 163tr PR/Pekka Parviainen 163b Marko Rikonen 164tl Photos.com/JI 164r Trudy E. Bell 164b Ken Crossland 165tr Wikipedia/Fir0002 165b Photos.com/JI 166tl SS/Catherine 166r PR/Frank Zullo 167tr PR/Gordon Garradd 167b Wikipedia/_64 168tl Wikipedia/Erik Axdahl 168r Photos.com/JI 169l Wikipedia/Lieutentant(j.g) 169r Wikipedia/Joseph N. Hall 170tl Trudy E. Bell 170c Pleum Chenaphun 170b Trudy E. Bell 171 NPS/Norbert Rosing

Chapter 11: Weather on Other Planets
172 NASA 173t NASA 173b NASA 174tl NASA 174r NASA/JPL/North 175tr NASA 175b NASA/Mark Robinson 176tl NASA/JPL 176b NASA 177t NASA/JPL 177b NASA/JPL 178tl NASA/JPL/MAIN 178c NASA/JPL 178b NASA/JPL/University of Arizona 179 NASA/JPL/MSSS 180tl NASA/JPL/University of Arizona 180c NASA/John Clarke/University of Michigan 180b NASA/JPL 181 NASA/ESA/E. Karkoschka/University of Arizona 182tl NASA/PR 182tr NASA 182br NASA/JPL/SSI 183t NASA/JPL/University of Colorado 183b Ken Crossland 184tl NASA 184tr NASA 184r NASA/L. Sromovsky and P. Fry/University of Madison-Wisconsin 185tr NASA/JPL 185b NASA/JPL

Chapter 12: Weather, Climate, and Society
186 AP/Frank Franklin II 187t SS/Cloki 187b AP/Ajit Solanki 188tl SS/Eric Limon 188tr NGS/Michael Lewis 188bl SS/Stephen Coburn 189 Photos.com/JI 190tl NGS/Maria Stenzel 190r NASA/KSC 191tr Ken Crossland 191b AP/David J. Phillip 192tl Public Domain 192bl USMA 192r LoC 193 AP/US Army 194tl Clipart/JI 194tr AP/Mary Evans 194r Robert A. Rhode 195 LoC/Harper's Weekly 196tl SS/Richard C. Bennett 196b AP/Vincent Yu 197t NASA 198tl SS/Paul Prescott 198tr Ken Crossland 198b PR/Yva Momatiuk and John Eastcott 199bl AP/Greg Baker 199c SS/Falk Kienas

At the Smithsonian
206 SI/Eric Long 207t SI/Dane A. Penland 207b NOAA/Mary Hollinger 208 NASA/EO

Cover
Front IO/Ablestock **Background** digitalvisions